Spacetimes Matter

A Collection of Mapping Methodologies

Jamie-Scott Baxter,
Anna Juliane Heinrich,
Séverine Marguin,
Vivien Sommer (Eds.)

jovis

To Gabriela Christmann
Your ongoing generosity and intellectual curiosity continue to inspire us.

SPACETIMES MATTER

Jamie-Scott Baxter,
Anna Juliane Heinrich,
Séverine Marguin,
Vivien Sommer (Eds.)

jovis

Jamie-Scott Baxter,
Anna Juliane Heinrich,
Séverine Marguin,
Vivien Sommer

Introduction

Multiple Spacetimes and How They Matter

In a world undergoing profound and rapid transformation, there is an urgent need to empirically grasp the complex dynamics of change. It is of paramount importance to understand the sociomaterial drivers for transformations, such as climate catastrophe, digitalisation, urbanisation, decolonisation, and the intersection of late capitalism with (neo)imperialism. These drivers exacerbate social and environmental inequalities, amounting to the radical destruction of democratic structures. We must learn how these forces operate on and reproduce local, regional, national, and supranational spatial formations, social practices, imaginaries, and subjectivities in deeply intertwined ways. To resist or effect change, we must critically examine the ways in which the very possibility of change is preconditioned through historical structural forces, entrenched power asymmetries, and hegemonic narratives. In short, we require sophisticated tools with an understanding of how pasts shape our presents; this is integral to grasping the complex, multiscalar, planetary processes that are continually reshaping the worlds we inhabit and the futures we can imagine.

One such approach can be found in what we call the 'refiguration of spaces' (Löw 2022; Löw and Knoblauch 2022; Knoblauch and Löw 2024). This theoretical programme calls for thinking about ongoing global changes through the analysis of space. As an alternative to merely juxtaposing globalisation and de-globalisation, the theoretical approach—grounded in an understanding of dynamic relational space—investigates multiscalar and non-linear disjunctures and conflicts around global phenomena. Using heuristics such as spatial figures (Löw 2020) and multiple spatialities (Knoblauch 2022), refiguration traces profound transformations that began in the nineteen-sixties and accelerated over the intervening decades. These shifts are marked by the expansion, extension, and intensification of circulation of people, goods, and information on increasingly large scales. These fundamental changes are experienced through and manifest in translocalisation (i.e., linking distanced spaces simultaneously) and polycontexturalisation (i.e., being embedded in several contexts at the same time), which are core concepts addressed by refiguration theory. This resonates with Doreen Massey's rejection of 'space as a static slice through time' (Massey 2005, 9). For Massey (2005), a multiplicity of spaces exists simultaneously within different temporalities. To fully grasp the ongoing refiguration of spaces, we argue that it is essential to theoretically and empirically consider different times—or what we call—multiple temporalities, and the ways in which they are entangled with multiple spatialities.

This book aims to capture and analyse the multiplicity of spacetimes. It presents a collection of mapping methods from colleagues and collaborators across different disciplines and parts of the world. Each contribution presents a novel methodology, developed through experimentation, for integrating multiple temporalities into spatial mapping. This results in a sequence of chapters, that are distinct and individual, but share a clear common purpose: providing critical reflections on methodological experiments. While some approaches emerged from or were tested during fieldwork, others have a more analytic or speculative purpose. Alongside methodological reflections, the authors draw out the multiple temporalities specific to their cases or projects. In this way, the volume offers at least two contributions. Firstly, it presents thirteen innovative mapping approaches that can be replicated and adapted to other sites, situations and objects of knowledge. Secondly, it assembles a collection of multiple spacetimes. Together, these objectives serve to deconstruct the hegemony of linear progressive clock-time that—all too often—shapes spatial research, transformation, and even the lived experiences and subjective knowledge of everyday life.

In this respect, the book may appeal to a broader audience across the sciences, especially those intellectually engaged in the so-called 'design turn' occurring in corners of the social sciences and humanities (Mareis 2010). More specifically, we expect the book and its collection of approaches to resonate with spatial researchers and designers, especially colleagues from architecture, planning, sociology of space, geography, and those engaged in transformation and transformative research. As the book's aims are focused and each essay tackles these joint objectives, we chose not to divide the book into subsections. Instead, each chapter presents a mapping protocol and is organised alphabetically by title. Authors were invited to coin a name for their approach, incorporating it into their essay titles. This organisation creates some unruly contrasts, chance encounters, and unexpected alliances as contributions line up next to one another. Regardless of order, the book is designed to be explored at your leisure—readers can dip in and engage with essays as they choose. Importantly, the book aims to strike a balance between text and images, ensuring that the visualisations are not merely illustrative but integral to the methodological discussions. Rather than serving as passive representations, the mappings function as analytical tools in their own right, serving to draw out and complement the written arguments. They do not simply depict spatial-temporal relations but actively shape how the authors understand and conceptualise them. Finally, we hope the book offers a tactile and aesthetic reading experience, inviting intellectual engagement and visual curiosity.

Mapping: All Space and No Time!

Situated in the design turn, the book emerges from an ongoing project to develop interpretative, qualitative and reflexive methods at the intersection of design and social scientific disciplines (Heinrich et al. 2024). It aims to engage a qualitative empirical socio-scientific approach, focused on the understanding of meaning-making through mapping. Thus, we try to bridge highly reflexive, theory-led social scientific approaches with designerly ways of knowing, based on synthetic drawings rooted in more associative and speculative thinking. These efforts resulted in what we call *hybrid mapping* (Baxter et al. 2025). Hybrid mapping is an interpretative method of analysis that visually synthesises heterogeneous data (e.g., quantitative/qualitative, multimodal), combining interdisciplinary ways of knowing from design practice and social science research. Initial operations include the spatial coding of qualitative interviews and ethnographic data into multiple drawings that relationally integrate spatial knowledge, spatial practices, and spatial structures. These are followed by an iterative and abductive composition process in which individual drawings are integrated into synthetic mappings to draw out the multiple spatial arrangements at work in these settings. Conceptual analysis unfolds by visually working out points of congruence, difference, and conflict complemented by textual elements, such as the development of legends and annotations in the mapping. This approach proved extremely fruitful for data collection and spatial analysis. Particularly, it allowed us to trace progressions in time through, for example, the incorporation of a sequence of spatial morphology drawings (i.e., a series of chronological maps). However, the method was less capable of capturing the complex temporalities that the theory of refiguration of spaces seeks to address. This provoked the question of how time, or more precisely, multiple temporalities, can be integrated into spatial mapping to allow us to capture refigurative processes.

To begin to answer this question, we organised two international online lecture series at the Technische Universität, Berlin in 2023 and 2024. These series, entitled Spacetime Matters, invited interdisciplinary speakers from Brazil, England, Germany, Switzerland, North America, France, Norway, the Netherlands, and Denmark. They presented and shared with an international audience a range of mapping experiments that aimed to integrate time into spatial research. The contributions presented novel mapping methods grounded in diverse contemporary theoretical programmes, including decolonial and Indigenous scholarship, critical and participative cartographies, feminist and new materialisms, more-than-human and feral ecologies, critical design and landscape theory, and refiguration of spaces. Papers reflected on at least two levels: firstly, how spacetime imaginaries are integrated into these theoretical programmes; secondly, on the ways in which mapping was being used to collect, analyse, and synthesise spatial and temporal data against these theoretical concerns. The contributions moved beyond theoretical spacetime speculations to consider how contested spacetimes play out and structure matters of fact and concern (Latour 2008) across scales in fields of migration, biodiversity conservation, rural urbanisation, geopolitics, landscape, urban studies, architecture, and memory studies. The papers, ideas, methods and discussions shared during the lecture series form the foundation of this book. As a group of inter- and transdisciplinary researchers, practitioners, and artists from diverse regions and time zones, we are convinced that *spacetimes matter.*

Spacetime Entanglements in Contemporary Thought

Theoretically, it has long been recognised that space and time should be considered together. Notable scholars have contributed to this understanding through various concepts. From a (neo)Marxist perspective, David Harvey (1990) explores the concept of 'spatiotemporality', focusing on how spatial and temporal dimensions are interconnected, particularly within the context of capitalism. He discusses how the acceleration of capital flows and the compression of space and time influence social and economic structures (Harvey 1990; Sheppard 2008). With his concept of spatiotemporal fix, Harvey aims to 'explore the forms and periodization of capitalist imperialism and to explain the overall logic of its latest, neoconservative phase' (Jessop 2006, 152). In a similar vein, Doreen Massey (2005) introduces 'space-times', arguing that space is not a static container but is intertwined with time and social relations. For Massey, space is a dynamic entity, constantly shaped by social interactions and imbued with stories and memories that weave together different temporalities. Building on Massey, Karen Barad (2007) proposes the term 'spacetimemattering' to emphasise the inseparability of space, time, and matter in the material-discursive making of worlds. Barad argues that these elements are entangled and co-constitutive. Her work on the atomic bombing of Hiroshima by the United States challenges these traditional distinctions by showing the enduring impact of atomic warfare on history, memory, and ecology (Barad 2017). Developing the integration of ecology in spacetime, scholars working on the Anthropocene have proven to be great sources of inspiration. Notably, the postcolonial scholar Dipesh Chakrabarty introduced the idea of 'planetary time', highlighting how human activities have become a geological force, altering the planet's systems and challenging traditional historical spatial narratives (Chakrabarty 2021). Similarly, Arturo Escobar advocates for decolonial transformative research and design for *transitioning* to the pluriverse. He emphasises the interconnection of social and ecological systems in transformation (Escobar 2015). From the perspective of refiguration theory, Gunter Weidenhaus's (2015) concept of 'social space-time' emphasises the deep interconnection between space and time in subjectivation processes. He highlights the importance of understanding the spatial and temporal dynamics that shape biographies and cultural shifts. Weidenhaus's approach encourages understanding space and time as active, co-constitutive elements in shaping individual and collective experiences and forms, which is a critical basis for the theory of refiguration of spaces.

Collectively, the cited scholars have contributed to the understanding that space and time are not separate entities but are deeply interconnected and co-constitutive, shaping social, economic, and environmental processes. However, while their theoretical insights are invaluable, they do not provide a concrete methodological framework for capturing, researching, and analysing the entanglements of spatial and temporal processes.

A Critical Look at Spacetime Methodologies

This book contributes to a growing discourse around mapping in spatial studies (Rankin 2016). On the one hand, the discourse takes a rather positivist approach by portraying mapping as a tool for cartographic measurement, for example by engaging GIS-software (Brunsdon and Comer 2019) or tools like space syntax (Psarra 2018). On the other hand, mapping discourses tend towards a more normative posture, pleading for critical participatory methods of investigation (Kollektiv Orangotango 2018) or design tools for projecting futures (Giseke et al. 2021; Schoonderbeek 2022). In these two currents, recent advancements in mapping approaches related to spacetime in the social sciences—particularly in architecture, planning, geography, cultural studies, and sociology—have emphasised the integration of spatial and temporal dimensions in analytical frameworks.

In the first, Geographic Information System (GIS) continue to be a foundational tool in capturing and quantitatively analysing social phenomena through spatial-temporal patterns. GIS-based mapping methods have been instrumental in studying crises such as the COVID-19 pandemic and in visualising temporal progression (Kianfar and Mesgari 2022; Ahasan et al. 2022). These findings underscore the potential of GIS in dynamically representing social processes over time, ranging from migration patterns and mobility studies to criminological hotspot analysis. In the second, interdisciplinary mapping techniques have emerged as a qualitative response to the limitations of traditional GIS. Font-Casaseca and Rodó-Zárate (2024) argue that conventional GIS models struggle to capture complex social relations, particularly regarding concepts of place, emotion, and multi-scalar spatial interactions. They propose integrating feminist, queer, and postcolonial perspectives into mapping methodologies to challenge dominant spatial narratives. Rose-Redwood et al. (2020) contribute to this discourse by exploring Indigenous mapping practices, which often conceptualise space and time relationally rather than in linear formats. Participative mapping methods have advanced significantly, for example story-mapping (Molden 2020), or geo-phenomenology (Iwai 2024). Such techniques enable a deeper analysis of spatial-temporal dynamics, integrating personal narratives with geospatial data in interactive maps.

Despite these significant advances, existing tools also reveal notable limitations. One shortcoming is their restricted ability to integrate highly diverse, multimodal datasets or time-based media which are essential for capturing lived and embodied experiences of spacetime. GIS-based methods offer little room for intuitive or abductive analytical processes, which we consider essential for understanding spatiotemporal entanglements. The subjective and affective dimensions of spacetime—how it is lived, experienced, and perceived—are difficult to capture within the rigid structures of conventional spatial analysis. Additionally, while some approaches aim for a relational understanding of space, they often remain constrained by a more rigid spatial framework. This is evident, for instance, in the way that social and material dimensions are sometimes treated as separate layers rather than as mutually constitutive forces. Furthermore, many tools remain heavily reliant on georeferencing, assuming that spatial analysis must always be tied to fixed geographic coordinates. From our perspective, space is not inherently defined by geolocation but by the dynamic relations and interactions that unfold. Additionally, the lack of multi-scalar approaches limits the ability to connect micro- and macro-level phenomena, from individual and more-than-human experiences to planetary-scale transformations. Perhaps most critically, these approaches tend to rely on singular conceptions of time, making it difficult to account for the multiple, coexisting temporalities that shape spatial formations. This is particularly problematic for postcolonial and feminist research, which challenges dominant Western notions of progressive time and calls for methodologies that can accommodate non-linear, cyclical, and ruptured temporal structures.

Against this backdrop, we build on theoretical insights into spacetime entanglements to push methodological developments further. This volume introduces a methodological spectrum that embraces a process-relational approach, allowing us to think beyond fixed spatial and temporal categories.

By incorporating multiple spatialities and multiple temporalities, we aim to develop a methodology that more adequately reflects the complexities of lived, experienced, and emergent *spacetimes*.

Spacetime Mapping

To consider how spacetimes matter, the rest of this introductory article and the following book make the case and provide trajectories of a methodological programme we call 'spacetime mapping'. We believe this overarching programme fills a methodological gap and links spacetime theory with empirical, qualitative, and interpretative research. We are also convinced that this approach connects design and research in effective ways, promoting transformative design research tasked with redesigning the material conditions for socio-spatial and ecological change. Spacetime mapping can be used as a diagnostic, analytical, or speculative-reflexive design tool.

As a continuation of the hybrid mapping method, spacetime mapping aims to bring into relation (multiple) spatialities and (multiple) temporalities. Spacetime mappings aim to draw out spatiotemporal situations and structures. They are particularly suitable for considering non-linear, non-hegemonic temporalities, such as circular time, seasonal time, deep time, geological time, time ruptures, etc. This can include the integration of time-based media (e.g., video or animations) into the multimodality of hybrid mapping. If we were to situate our approach within broader epistemological shifts, it would resonate most with the processual and material turns occurring in the social sciences. These perspectives move beyond static representations of space and linear understandings of time, emphasising relational, emergent, and multi-scalar processes while not losing sight of political and affective matters.

In the remainder of this chapter, we will introduce the individual contributions and their key insights before revisiting the central question (what is spacetime mapping?) in the short concluding section.

Multiple Spacetime Mappings

In the first contribution, Ana Luiza Nobre and David Sperling introduce their web based counter-cartography project, the *Ground Atlas*. For them, 'Decolonial Mapping' means making visible conflictual historical urbanisation processes and re- and deterritorialisation, which are inscribed in the ground today. This is done through a critical reinterpretation of historical events, demonstrating how past, present, and future are entangled rather than sequential. Rejecting linear, Eurocentric conceptions of time, the authors introduce 'spiral time' as an alternative temporality grounded in African and Afro-diasporic cosmologies. This joins up with a common call that connects thinkers, practitioners, and activists across the planet. Nobre and Sperling advocate for an approach to mapping grounds based on an 'ethic of life' that acknowledges the interdependency between all beings and entities (Nobre and Sperling this volume, 21).

The second article, similarly acknowledging the radical interdependency of the world, challenges static spatial cartography in other ways. Jamie-Scott Baxter and Vivien Sommer's 'Diffractive Mapping' puts forward a method for analysing *spatiotemporal phenomena* to consider how matters of space and time are co-constituted in material-discursive practices (Barad 2007). Rooted in Barad's agential realism, Baxter and Sommer introduce an interdisciplinary method at the juncture of sociology and architecture that traces events, processes, and intra/interactions. Theoretically situated in the materialist turn, their approach recognises the processual nature of spatial matters and offers a reflexive and systematic mapping method through which to analyse social change and spatial transformation.

Jae-Young Lee's 'Experiential Mapping' offers an autoethnographic method for spatial research. In the chapter, Lee examines the entanglement of subjective and nonhuman materialities in spatial knowledge production. The method was constructed during her empirical research in South Korea, as a way to reflect on her positionality in the field. Through daily mapping, sketches, interviews, and field notes, she considers how her historically situated spatial knowledge impacts her spatial research. The deeply reflexive methodology brought to the surface three temporalities that shaped the fieldwork, including the monsoon season, national vacation season, and post-vacation summer. The contribution highlights the complex interplay between a researcher's subjectivity, spatial perception, and spatial research while furthering an understanding of how time shapes knowledge production.

Resonating with Nobre and Sperling's *Ground Atlas* project, Lili Carr, Alder Keleman Saxena, Victoria Baskin Coffey, Jennifer Deger, Feifei Zhou, and Anna L. Tsing present their web-based *Feral Atlas*. The chapter describes how what they call 'Flow Mapping' captures the more-than-human processes of the Anthropocene, demonstrated in the Atlas. Their approach is grounded in the more-than-human turn in the social sciences, which recognises the 'co-mingling' of human and nonhuman lifeways (Carr et al. this volume, 65). Similarly to Baxter and Sommer, they take a process-relational approach to mapping in order to capture flows, events, and ecological dynamics in the transformation of landscape structures (Tsing et al. 2019). They illustrate this with flow maps of frogs, pathogens, invasive plants, and disease which reveal the spatiotemporal 'feral dynamics' of leakages, spills, blockages, and simultaneities co-constituting the Anthropocene.

In the next chapter, Anne-Marie Willis and Tony Fry revisit Willis's groundbreaking article from 2012, 'The Ontological Designing of Mapping', which argued that maps are ontological designing forces acting upon the world, while simultaneously recognising that the 'world designs back' (Willis 2012). With the contemporary example of 'Future-risk Mapping' in climate change, they update their argument to show how maps, while still 'ontologically formative', mark 'ending the time of one thing and inaugurating the time of what is to come' (Willis and Fry this volume, 79). They go on to argue that risk-mapping is insufficient to capture the complexity and meaning of socioecological transformations. Instead, they discuss the idea of 'designing in time', which requires designers to take a speculative spacetime perspective. For designers to (pre)consider the potential consequences of what they design involves 'designing back from the future' (*idem*). This has radical consequences for spatial designers, challenging conventional tools such as the site plan and site analysis which, until now, have tended to only consider the histories and present conditions of sites.

Angela Million, Simten Önen, and Katrin Schamum illustrate 'Gamified Mapping' as a tool for ethically investigating the well-being of children in spacetime. The chapter presents a study grounded in Massey's conceptualisation of spacetime as the 'simultaneity of stories-so-far' (Massey 2005, 9). It develops a gamified approach to participatory mapping in a Berlin children's centre, where children's spacetime knowledge, their narratives, and perceptions are collectively mapped. The authors end by reflecting on the ethical dimension of doing participative spatial research with vulnerable groups, especially considering how to ensure anonymity when collecting spatiotemporal narratives in maps—which may risk revealing locational and biographical information.

In his chapter, Qusay Amer explores the interrelation of space and time in migration and refugee movements in Amman. Building on Weidenhaus's (2015) 'social space-time' concept, Amer examines how migration flows shape urban spaces in overlapping, recursive patterns. Using 'Historical-Network Mapping', the study reveals how governance, economic forces, and spatial negotiations influence refugee settlement. By challenging linear narratives of urban development, the chapter advocates for multi-scalar, non-linear spacetime mapping methods that better capture migration's volatility, simultaneity, and political dimensions, thus offering a more nuanced understanding of urban transformation and refugee experiences.

Chiming with *Feral Atlas* and *Ground Atlas*, Jamie-Scott Baxter, Lus Constantin, and Séverine Marguin present their web-based 3D interactive space, the 'Botaniverse', an example of what they call 'Immersive Mapping'. Experimenting with the possibility of VR technology in the metaverse, the authors aimed to integrate time-based media, such as animation, video, audio, etc., into spatial mapping. To do this, they de- and re-constructed spatial maps they had previously made with students and extracted four key narratives that set a stage for a new digital botanic garden. These were: colonial legacies, contested knowledge systems, plant agencies, and the digitisation of archives. The method served to review and reinterpret spatial data collected during a three-year research project on botanical gardens. It yielded a new interpretation of how multiple overlapping spacetimes structure the refiguration of Berlin's botanical gardens. These were colonial, planetary, and perpetual spacetimes, showing how past and present knowledge systems shape botanical gardens and challenge a linear view of globalisation.

Noël van Dooren explores time-space integration in landscape design, distinguishing between observational and exploratory drawings. He builds on the Long Now Foundation's temporal framework—'these days', 'our times', and 'the long now'—to articulate the complexities of representing time in design. Extending the scope to include historical precedents like Humphry Repton's 'before and after' technique, van Dooren presents 'Long-Now Mapping' to capture landscape transformations. In his contribution, he examines how landscape drawings can draw out multiple temporal scales in different ways. Echoing Willis and Fry in this volume, he concludes by calling for the integration of temporal dimensions into spatial design to enhance the design of future dynamic and responsive landscape systems.

With her chapter about 'Mapping Fleetingness', Francesca Ceola advocates for spacetime-sensitive mapping that captures the volatility of displacement, fluidity and uncertainty. The approach highlights how urban pressures incrementally erase internally displaced persons' settlements in Lagos. It underscores the ethical and political dimensions of mapping, emphasising participatory approaches that empower displaced communities to assert their spatial rights and document their spatial agency. Her chapter concludes with Ceola reconceptualising forced displacement dynamics as non-linear and cyclical rather than temporary.

Philippe Rekacewicz's text discusses the complexity of representing temporal evolution in maps. It juxtaposes synchronic (snapshot in time) and diachronic (evolution over time) perspectives, referencing linguistic theories by Ferdinand de Saussure. The text introduces 'Multidimensional Cartography' to include additional dimensions, such as verticality (above and below ground) and the imaginary. The method divides long historical periods into key moments (e.g., European border changes in the twentieth century) to make complex temporal dynamics more comprehensible. It incorporates human emotions, perceptions, and personal experiences of spacetime, advocating for maps that reflect lived realities rather than abstract statistics.

Simone Rueß, an artist, employs a performative and embodied approach to investigate how historical, economic, and social transformations are reflected in everyday life. Using a mobile mapping station in postsocialist Kraków, she engages residents through participatory arts-based research. Her 'Performative Drawing' method captures the tactile experience of urban space; like Baxter, Constantin, and Marguin's 'Botaniverse', it is later processed into animations and an immersive installation. This work challenges the boundaries between mapping, artistic intervention, and spatial research, revealing the layered biographies embedded in urban environments.

Finally, Dagmar Pelger's contribution 'Time-Deep Mapping' examines urbanisation through the lens of the commons. It challenges dominant historical urban development narratives, resonating with Nobre and Sperling's critical cartography. Pelgar shows how emancipatory and civil society movements have shaped the city, but are often erased by powerful hegemonic spacetime stories, formed and perpetuated by capital. She traces historical contestations and enclosures as well as the reclamation of common spaces, squatting movements etc., showing the non-linear, recursive, and (dis)continuous trajectories of urban ownership models, oscillating between public and private, common, and club. Pelger's 'Time-Deep Mapping' serves as both an analytical and political tool, reframing urban space as a site of continuous struggle between enclosure and collectivisation, private and common interest.

What is Spacetime Mapping?

Conventional maps all too often depict space as singular while concealing time altogether. With this in mind, the collection of mapping methods summarised above and presented in the following chapters challenge hegemonic conceptions of 'space as a static slice through time' (Massey 2005, 9). Collectively, they provide multiple examples of what we call *spacetime mapping,* offering new ways of understanding and capturing complex spatiotemporal entanglements. The final paragraphs of this editorial aim to synthesise key aspects of these approaches and draw out four dimensions central to the spacetime mapping methodology. These are: *processual, political, imaginative, and affective.*

Processual Dimension

Grounded in process, or process-relational ontology (see e.g., Bergson 1907; Whitehead 1929; Deleuze 1994), spacetime mapping foregrounds relations, processes, and interwoven social and material dynamics. Spacetime mapping draws in/out/on multiple spacetimes. It tracks and traces material-discursive flows, more-than-human dynamics, and spatiotemporal events to reveal recursive socio-spatial patterns and polycontextural overlays and overlaps (Amer; Baxter and Sommer; Baxter, Constantin and Marguin; Ceola this volume). The approach recognises the radical interdependence between beings. It acknowledges that spacetime worlds are not static and inert but continually emerging in the interaction between human and nonhuman forces (Nobre and Sperling; Carr et al. this volume).

Political Dimension

Map-making is not a neutral practice, it is also an ontological and performative force in the becoming of worlds (Rueß; Willis and Fry this volume). This makes mapping spacetimes a political ontology, where what is at stake is not only knowledge of the world but what and how worlds come to exist (Blaser 2013). Mapping the multiplicity of spacetimes is a reality-making project in which worlds are enacted. While the power of maps to make worlds has long been critiqued in critical cartography discourses, spacetime mapping aims to reflexively harness this potential for socially and environmentally just transformation. This involves making visible *minor spacetimes* that have been suppressed, erased, and overwritten by hegemonic actors and processes (Pelger; Nobre and Sperling this volume). Or, drawing in the spacetime knowledge of vulnerable groups, including children (Million et al. this volume), migrants and refugees (Amer this volume), and internally displaces people (Ceola this volume). Yet, as Pelger cautions, this is a struggle in which subaltern spacetime stories need to be retold constantly in order to continually produce them. Where minor spacetimes can be enacted through mapping, hegemonic *major spacetimes* can be deconstructed. This reveals the ways in which spacetimes are structured by power dynamics such as colonialism (Baxter, Constantin and Marguin; Nobre and Sperling this volume), patriarchy and capital (Pelger this volume), war (Rekacewicz this volume), and displacement (Ceola; Amer this volume).

Imaginative Dimension

To consider the relations in and between multiple spatialities and multiple temporalities, new spacetime imaginaries are required. This means finding not only new ways of conceptualising futures (van Dooren this volume), but also an understanding of how pasts, presents, and futures tangle—disrupting linear progression. Approaches like 'designing back from the future' (Willis and Fry this volume) challenge conventional temporal assumptions, offering alternative ways of envisioning what is yet to come. However, for this not to become an indulgent intellectual exercise or to fall into the trappings of conventional cartography, it is paramount for designers and researchers to continually reflect on their own knowledge, imaginaries, and positional-ities and consider how these affect the object of their endeavours, as Lee's auto-ethnographic account so well illustrates.

Affective Dimension

Spacetime mapping is not only a cognitive or analytical practice but also an affective, sensory, and immersive experience (Baxter, Constantin and Marguin; Nobre and Sperling this volume). Spaces and times are not merely shaped by abstract relations or political power structures but are also lived through emotions, memories, atmospheres, and bodily sensations (Rueß this volume). This dimension explores how mapping methods can capture not just material or social processes but also moods, movements, and sensory experiences. By integrating this dimension, spacetime mapping expands beyond conventional cartographic practices to embrace the ways in which space and time are felt as much as they are structured (Rekacewicz; Lee this volume).

These reflections are meant as a starting point, an invitation to engage and develop the multiplicity of spacetime mappings further. We hope you will enjoy reading the following contributions as much as we have enjoyed working together on this collection and the ideas within.

Acknowledgements
We would like to extend our appreciation to the DFG-funded collaborative research center 'Refiguration of Spaces' (CRC 1265) for its ongoing support in organising the lecture series and realising the book. Especially Nina Elsemann for her enthusiasm and trust. Many thanks to Mara Oppermann for her great coordination and to Katja Gretzinger for her fantastic design. It was a delight to work with you all! And of course, thank you to all our contributors: it is an honour to have gathered your wonderful work into this anthology.

References

Ahasan, Rakibul, Md. Shaharier Alam, Torit Chakraborty, and Md. Mahbub Hossain. 2022. 'Applications of GIS and geospatial analyses in COVID-19 research: A systematic review', *F1000 Research*, 9(1379): pp. 1–20. doi:10.12688/f100research.27544.2.

Barad, Karen. 2007. *Meeting the Universe Halfway: Quantum Physics and the Entanglement of Matter and Meaning*. Durham: Duke University Press.

Barad, Karen. 2017. 'No Small Matter: Mushroom Clouds, Ecologies of Nothingness, and Strange Topologies of SpaceTimeMattering'. In *Arts of Living on a Damaged Planet: Ghosts and Monsters of the Anthropocene*, edited by Anna Tsing, Heather Swanson, Elaine Gan and Nils Bubandt: pp. 103–120. Minneapolis: University of Minnesota Press.

Baxter, Jamie-Scott, Séverine Marguin, Sophie Mélix, Martin Schinagl, Ajit Singh and Vivien Sommer. 2025. 'For Hybrid Mapping: A Methodology to Study Relational Space. The Case of Botanical Gardens Berlin'. *Architecture and Culture* 11(3).

Bergson, Henri-Louis. 2019 [1911]. *Creative Evolution*. New York: Gray Rabbit Publishing.

Blaser, Mario. 2013. 'Ontological Conflicts and the Stories of Peoples in Spite of Europe: Toward a Conversation on Political Ontology'. *Current Anthropology* 54(5): pp. 547–568.

Brunsdon, Chris, and Lex Comber. 2019. *An Introduction to R for Spatial Analysis and Mapping*. London: SAGE Publications Ltd.

Chakrabarty, Dipesh. 2021. *The Climate of History in a Planetary Age*. Chicago: The University of Chicago Press.

Deleuze, Gilles. 1994. *Difference and Repetition.* New York: Columbia University Press.

Escobar, Arturo. 2015. 'Transiciones: a space for research and design for transitions to the pluriverse'. *Design Philosophy Papers* 13(1): pp. 13–23.

Fagerholm, Nora, Christopher M. Raymond, Anton Stahl Olafsson, Gregory Brown, Tiina Rinne, Kamyar Hasanzadeh, Anna Broberg, and Marketta Kyttä. 2021. 'A Methodological Framework for Analysis of Participatory Mapping Data in Research, Planning, and Management'. *International Journal of Geographical Information Science* 35(9): pp. 1848–75. doi:10.1080/13658816.2020.1869747.

Font-Casaseca, Nuria and Maria Rodó-Zárate. 2024. 'From the margins of Geographical Information Systems: Limitations, challenges, and proposals'. *Progress in Human Geography* 48(4): pp. 421–436. doi:10.1177/03091325241240231.

Giseke, Undine, Martina Löw, Angela Million, Philipp Misselwitz, and Jörg Stollmann, eds. 2021. *Urban Design Methods: Integrated Urban Research Tools*. Berlin: JOVIS.

Harvey, David. 1990. *The Condition of Postmodernity*. Cambridge: Blackwell Publishers.

Heinrich, Anna Juliane, Séverine Marguin, Angela Million, and Jörg Stollmann, eds. 2024. *Handbook of Qualitative and Visual Methods in Spatial Research*. Bielefeld: transcript Verlag.

Iwai, Yuki. 2024. 'Geo-Phenomenology: A Qualitative and Humanistic GIS Approach to Exploring Lived Experience'. *The Professional Geographer* 76(5): pp. 607–19. doi:10.1080/00330124.2024.2372794.

Jessop, Bob. 2006. 'Spatial Fixes, Temporal Fixes and Spatio-Temporal Fixes'. In *David Harvey: A Critical Reader,* edited by Noel Castree and Derek Gregory, pp. 142–166. Cambridge: Blackwell Publishers.

Kianfar, Nina and Mohammad Saadi Mesgari. 2022. 'GIS-based spatio-temporal analysis and modeling of COVID-19 incidence rates in Europe'. *Spatial and Spatio-temporal Epidemiology* 41: p. 100498. doi: 10.1016/j.sste.2022.100498.

kollektiv orangotango, ed. 2018. *This Is Not an Atlas. A Global Collection of Counter-Cartographies*. Berlin: transcript Verlag.

Knoblauch, Hubert. 2022. 'Multiple Säkularität, Multiple Räumlichkeit Oder Multipolarität – Auch Eine Kritik Der Postkolonialen Vernunft'. In *Verstehen Als Zugang Zur Welt. Soziologische Perspektiven*, edited by Uta Karstein, Marian Burchardt, and Thomas Schmidt-Lux, pp. 99–118. Frankfurt Main: Campus Verlag.

Knoblauch, Hubert and Martina Löw. 2024. 'The Refiguration of the Global: Globalization and the Spatial Logics of Digitalization'. *Critical Sociology* (online first). https://doi.org/10.1177/08969205241262304.

Latour, Bruno. 2008. *What is the Style of Matters of Concern?* Assen: Van Gorcum.

Löw, Martina. 2020. 'In welchen Räumen leben wir? Eine raumsoziologisch und kommunikativ konstruktivistische Bestimmung der Raumfiguren Territorialraum, Bahnenraum, Netzwerkraum und Ort'. In *Grenzen der Kommunikation. Kommunikation an den Grenzen,* edited by Jo Reichertz, pp. 149–164. Weilerswist: Velbrück.

Löw, Martina. 2022. 'Sozialen Wandel verstehen: Refiguration'. In *Verstehen Als Zugang Zur Welt. Soziologische Perspektiven*, edited by Uta Karstein, Marian Burchardt, and Thomas Schmidt-Lux, pp. 81–97. Frankfurt Main: Campus Verlag.

Löw, Martina, and Hubert Knoblauch. 2022. 'The Refiguration of Space, Circulation, and Mobility'. In *Spatial Transformations. Kaleidoscopic Perspectives on the Refiguration of Space*, edited by Angela Million, Christian Haid, Ignacio Castillo Ulloa, and Nina Baur, pp. 17–27. London: Routledge.

Mareis, Claudia. 2010. 'Entwerfen – Wissen – Produzieren. Designforschung im Anwendungskontext'. In *Entwerfen – Wissen – Produzieren. Designforschung im Anwendungskontext*, edited by Claudia Mareis, Gesche Joost, and Kora Kimpel, pp. 9–32. Bielefeld: transcript Verlag. doi:10.14361/transcript.9783839414637.9.

Martínez-Hernández, Carlos, Radosław Piskorski, and Arie Stoffelen. 2023. 'Designing Online Workshops for Teacher Trainees: Heritage Mapping With Web GIS Story Maps'. *European Journal of Geography* 14(3): pp. 68–78. https://doi.org/10.48088/ejg.c.mar.14.3.068.078.

Massey, Doreen. 2005. *For Space*. Thousand Oaks: SAGE Publications.

Molden, Olivia C. 2020. 'Short Take: Story-mapping experiences'. *Field Methods* 32(2): pp. 131–139. doi:10.1177/1525822X19877381.

Psarra, Sophie. 2018. *The Venice Variations*. London: UCL Press.

Rankin, William. 2016. *After the Map. Cartography, Navigation, and the Transformation of Territory in the Twentieth Century.* Chicago: The University of Chicago Press.

Rose-Redwood, Reuben, Natchee Blu Barnd, Annita Hetoevėhotohke'e Lucchesi, Sharon Dias, and Wil Patrick. 2020. 'Decolonizing the Map: Recentering Indigenous Mappings'. *Cartographica* 55:3: pp. 151–162.

Schoonderbeek, Marc. 2022. *Mapping in Architectural Discourse: Place-Time Discontinuities. Architectural Borders and Territories.* New York: Routledge.

Sheppard, Eric. 2008. 'David Harvey and Dialectical Space-Time'. In *David Harvey: A Critical Reader*, edited by Noel Castree and Derek Gregory, pp. 121–141. Cambridge: Blackwell Publishers.

Tsing, Anna L., Heather Swanson, Elaine Gan and Nils Bubandt. 2019. *Arts of Living on a Damaged Planet: Ghosts and Monsters of the Anthropocene*. Minneapolis: University of Minnesota Press.

Weidenhaus, Gunter. 2015. *Soziale Raumzeit.* Berlin: Suhrkamp Verlag.

Whitehead, Alfred. 1929. *Process and Reality: An Essay in Cosmology.* New York: Macmillan.

Willis, Anne-Marie. 2012. 'The Ontological Designing of Mapping'. *Design Philosophy Papers.* 10(2): pp. 95–104.

Ana Luiza Nobre,
David Sperling

Decolonial Mapping

The *Ground Atlas*

'Where to land?' (Latour 2018). This question, raised by Bruno Latour a couple of years before the COVID-19 pandemic, paved the way for the *Ground Atlas* project [groundatlas.org]. The *Ground Atlas* is a web-based countercartographic and historiographic experiment, ongoing since 2020. It was established in partnership between the Laboratory of Architectural Analysis in the Department of Architecture and Urbanism at the Pontifical Catholic University of Rio de Janeiro and the Center for Studies on Contemporary Spatialities in the Institute of Architecture and Urbanism at the University of São Paulo, with support from FAPERJ and the Brazilian National Council for Scientific and Technological Development[1].

1 Conception and coordination: Ana Luiza Nobre and David Sperling. Design: Claudio Bueno and Vitor Cesar. Programming: Jeff Sobral. Team: Ana Carolina de Paula Bezerra, Alexandre Silveira, Caio Rechuem, Daniel Lavinas, Giovanni Bussaglia, Joana Martins, Mariane Cardoso de Santana, Michel Zalis, and Victor Vaccari.

The *Ground Atlas* was designed simultaneously as an open set of maps, an archive, and a constellation system of images (Figure 1, random homepage). It aims to map and make critically visible the complex processes of urbanisation, territorialisation, and deterritorialisation that remain somehow inscribed on the ground today in different historical-cultural and geopolitical contexts (not necessarily produced but somehow legible today, like 'footprints' left by human action over time).

The project takes Latour's question as a wake-up call to 'come down to earth' and ask ourselves how and with what/whom we want to live/inhabit/build worlds on this planet, considering the other species that cohabit it with us at this dizzying moment 'when the ground is giving way beneath every one's feet' (Latour 2018, 8). At this cosmic crossroads, the voices of Indigenous and Quilombola leaders, such as Ailton Krenak (2020), Davi Kopenawa (2010), Jerá Guarani (2022), and Nego Bispo (2023), resonate stronger, along with those of academic thinkers, such as Anna L. Tsing (2015), Donna Haraway (2016), Rita Segato (2022), Eduardo Viveiros de Castro (2013), Malcom Ferdinand (2022), Maria Puig de la Bellacasa (2015), Vandana Shiva (2015), Achille Mbembe (2023), and Paulo Tavares (2020), in a common call for an ethic of life based on respect for the Earth and the interdependence between all beings and entities.

This implies a commitment to problematising worldviews that are linked to geographical coordinates as well as visual, two-dimensional representations of the planet we have inherited from European rationalism and that have profoundly conditioned our way of seeing, thinking, and acting. Images such as *Mercator's world map* (1569) and ***the** Blue Marble* (1972) correspond to an exclusionary worldview and paradigm of knowledge that were invented by modern science and based on the separation between subject and object and on supposedly reliable, 'neutral' representations of a pre-existing reality. These images need to be held in mind to understand the growing effort to set new worlds into motion through other procedures, paradigms, and imageries. Cartography undoubtedly plays a key role in this movement —not as an objective science dedicated to producing representations of the world but as a political-philosophical problem linked to the construction and configuration of more-than-human worlds and territorialities (Sperling 2024).

'Decolonization must begin on the ground'[2]

2 Alves, Maria T. 2022. *Seeds of Change: New York, A Botany of Colonization, 2017–2018*, p. 149.

At the intersection of space, time, narrative, and representation, cartography offers a common base to explore contemporary processes of spatialisation, urbanisation, and territorialisation that are deeply rooted in colonialism through an interdisciplinary and critical lens.

Therefore, the *Ground Atlas* assumes the challenge of exploring different ways of communicating (textual, visual, and intermediary) and fostering the configuration of worlds that are in line with the so-called decolonial epistemic turn unleashed by Latin American intellectuals, such as Walter Mignolo, Arturo Escobar, and Aníbal Quijano (Segato 2022). These worlds are also attuned with what Argentine sociologist Maristella Svampa (2019) has more recently defined as the 'ecoterritorial turn'. In other words, we stand alongside growing political-ideological movements across Latin America over the last two decades. Standing at the intersection of the struggles of oppressed

Figure 1 *Ground Atlas* platform (Ana Luiza Nobre and David Sperling, 2022)

and racialised peoples and feminist and ecological movements, these activists advocate in defence of land and territories seriously impacted by new forms of extractivism, exploitation, and socio-environmental injustice.

Inspired by the work of Brazilian artist and activist Maria Thereza Alves and her call to consider the ground as the starting point of decolonisation, we decided to root the *Ground Atlas* in a set of questions:

> How do we challenge and decolonise the consolidated narratives and apparatuses that have shaped urban historiography so far, considering, for example, the complexity of urbanisation processes in Ibero-American countries?
>
> How do we rearticulate and reveal complex and profound relationships between urban infrastructure works, energy sources, political discourses, power relations, slavery, migratory movements, social practices, ways of life, and struggles?
>
> How do we critically consider visible, complex processes—marked by conflicts, contradictions, and ambiguities—that involve urban spaces and, at the same time, go far beyond their limits?
>
> How do we free the imagination from cartographical and historiographical conventions and foster new political, urban, and planetary imageries?

Key to this project is the polysemous dimension and the unclear delineation of the word 'ground', which resists any univocal or pre-established scientific definition. We actually think of *ground* in many ways: as a world archive, in which all actions (anthropic and nonanthropic) are somehow inscribed and leave traces; as a living being, made up of the continuous interaction between multiple species; as a world, in which countless organisms, agents, geopolitical forces, and logics intersect; as a composite of multiple many times, ambivalent historical, political, biological, social, cultural, economic, and symbolic layers across scales and in different historical-cultural and geopolitical contexts; and as an essential condition for coexistence on a planet populated by a myriad of plants, animals, bacteria, and other life forms.

Attentive to the multivalence of meanings associated with the ground—whether in scientific, poetic, and/or mythical terms—the *Ground Atlas* aligns itself with other recent initiatives and projects in different parts of the planet. In focusing on practices of care and regeneration in its semantic preoccupation with soil/earth/land/terrain/territory—terms so implicated in each other that, in several languages, a single word points simultaneously to both the ground and the planet that we inhabit.

'It's beginning, middle and beginning'[3]

3 Bispo dos Santos, Antônio. 2023. *A Terra dá, a Terra quer.* São Paulo: Ubu/Piseagrama.

To assume a decolonial approach in cartography means considering both space and time as complex social and historical constructions to be interrogated from the perspective of the 'coloniality of power', as addressed by Quijano (2005)—in other words, by 'tensioning' the Eurocentric space-time conception that has established itself as a global standard since the conquest and constitution of America. This means looking critically not only at Mercator's projection and its derivatives but also at linear representations of time that sustain an orderly arrangement between past, present, and future in correspondence with the Euro-Christian cosmology that has fed, financed, and shaped persistent systems of extraction/colonisation.

In this sense, neither a territory nor time in itself can be thought of as something absolute, universal, measurable, and representable according to a straight line. In contrast to the axiality of Western cumulative and progressive time, contemporary thinkers such as Leda Maria Martins and Nego Bispo

Figure 2 Point interfaces (Ana Luiza Nobre and David Sperling, 2022)
Figure 3 Manto Tupinambá point interface (Ana Luiza Nobre and David Sperling, 2022)

have already pointed out African and Afro-diasporic cosmologies that conceive time as a spiral. In recent years, this curvilinear and ancestral temporality has become key in contesting Western logocentric thinking precisely because it dismisses the idea of time as an arrow that always points forward. After all, in spiral time, there is no beginning, middle, or end. Past and future exist, but not within rigid boundaries or as causes and consequences. Moreover, the past is by no means immutable.

According to Martins (2021), spiral time is 'foldable'—that is, it is reversible and does not necessarily move forward but rather in 'twists' or *ritornellos* that manifest themselves in the present, bringing with them past and future in movements of retrospection, prospection, contraction, and dilation that entwine all times. This is reflected in the well-known Yoruba saying, 'Exu killed a bird yesterday with the stone he threw today.'

'Isn't the atlas... the space for the greatest reserve of imagination?'[4]

4 Didi-Huberman, Georges. 2010. *Atlas ¿Cómo Llevar el Mundo a Cuestas?* Madrid: Museo Nacional Centro de Arte Reina Sofia: p.52.

In terms of spatial and narrative form, the *Ground Atlas* is based on two disruptive European thinkers: Aby Warburg (2010) and his *Mnemosyne Atlas* (along with its readings by Georges Didi-Huberman [2010]), and Walter Benjamin (2007) and his fragmentary writing. As key references to subverting traditional narratives, their works are mobilised in order to bring together—through a digital interface—images, texts, discourses, and logics within a transversal perspective that seeks to subvert all linear order and challenge canonical forms of historiographical writing. This can also be seen to create a 'reading machine' (Didi-Huberman 2010) that explores erratic paths, explodes conventional narrative forms, and invests in the multiplicity, anachronism, and hybridity inherent to the assembly operation.

The *Warburgian Atlas* is taken simultaneously as a countercartography and a counternarrative that challenges us to 'think by images'—in other words, through a form of reading radically different from textual and linear ones. This 'form atlas' was intertwined with Benjamin's constellation writing and with Gilles Deleuze and Félix Guattari's (1995) rhizomes. At the same time, we have investigated artistic, architectural, and cartographic practices that somehow fall within this lineage to feed creative processes and maintain a critical dimension that is, to some extent, similar to the ways in which knowledge power is constituted and often maintained.

In Benjamin's conception of history, we sought a kind of visual and spatial writing based on assembly, correspondence, and free articulation among loose points, fragments, and quotations that, at some point, form a drawing or diagram. Benjamin reminds us that constellations are not natural formations but cultural images projected onto the arrangement of the stars, with different configurations at different times. In this sense, it is up to the reader to 'contemplate' the texts and see, in the manner of a stargazer, the unexpected connections between different points and activate them mutually—hence, the intention to forward the experimentation already tested in previous works[5] of constellation writing (Figure 2).

5 See www.rionow.org and www.iau.usp.br/pesquisa/grupos/nec/wp-content/uploads/2017/06/ebook-GRU-111-Contracartografias-NEC-USP.pdf

6 'We don't own the land ... we are the land.' Casé Angatu at https://www.atlasdochao.org/ponto/yme-iande-jara-sui-yby-iande-yby/

In other words, rather than visually representing something already put into words, we try to give images a voice and thereby pave the way for thinking —and writing—through images and 'producing shock', in the Benjaminian sense, in order to 'make the past jump' out of the historical continuum. One of the great challenges, once again with Benjamin, is to intrigue rather than explain—that is, to allow oneself to be pressed by questions relating to how to write history, how to organise historical knowledge, how to mobilise images in the context of writing, and how to relate text and images.

'Ymé Îandê Jara Suí Yby ... Îandê Yby'[6]

The *Ground Atlas* also stems from a turn in cartography significantly related to the social history of maps. This means working against the supposed neutrality and impartiality of maps and looking at them as a form of language that

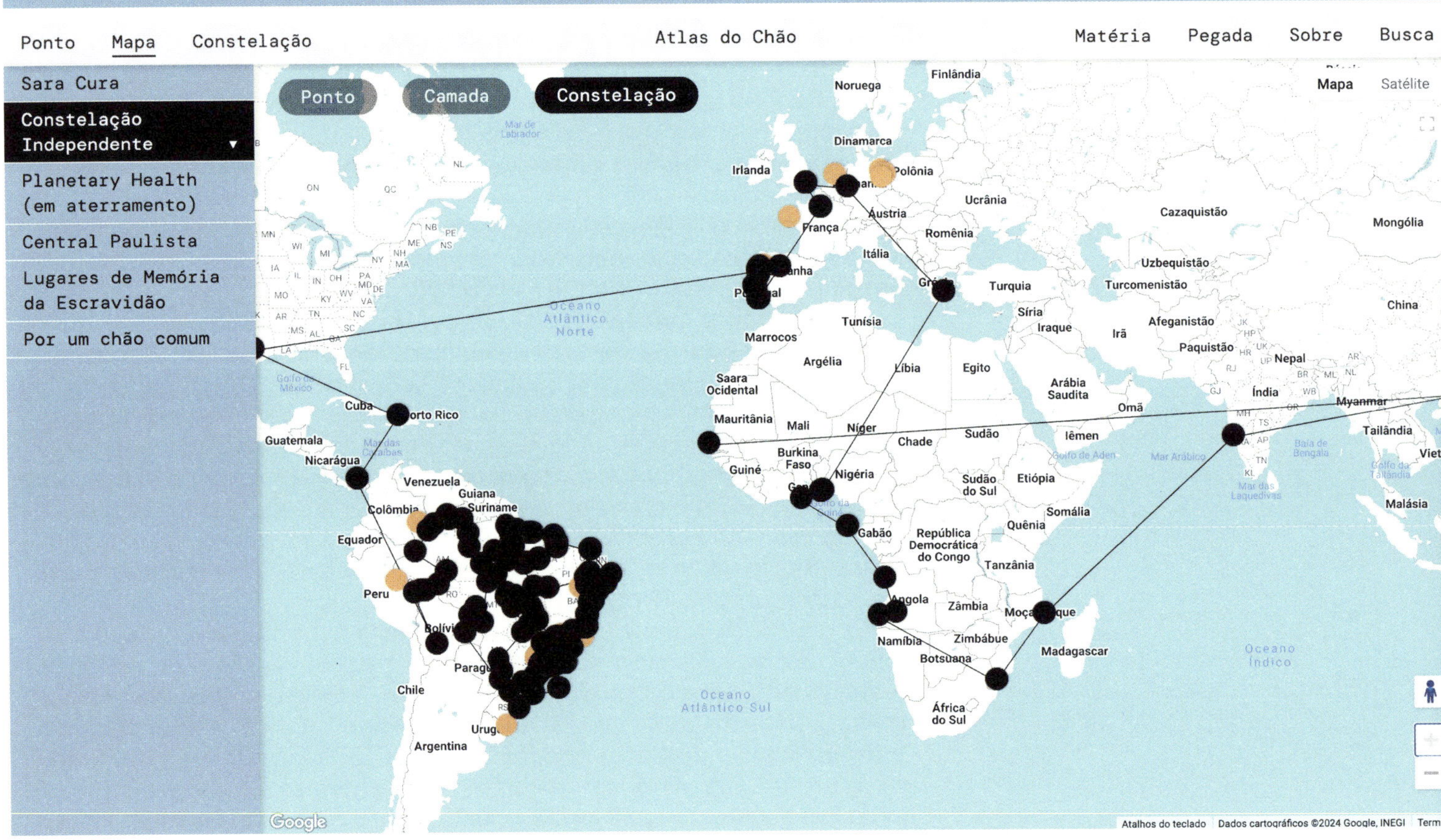

Figure 4 Constellations interface (Ana Luiza Nobre and David Sperling, 2022)
Figure 5 Independent Constellation Map interface (Ana Luiza Nobre and David Sperling, 2022)

belongs to the category of images—more specifically, socially constructed images, never free of judgement and neither true nor false. In this sense, the *Ground Atlas* places itself within a field that benefits significantly from the incorporation of two types of displacement. One is the ontological critique of scientific stances, according to which maps are assumed to be objective devices, mirrors of reality. This means putting the ways in which maps are made and their contents into dispute; unveiling power relations, ideologies, and historical determinations; and problematising their codes and referents. The other is a dialogue with a universe of open cartographic experiments regarding their political and creative aspects and different fields of knowledge and practice.

From a decolonial perspective, the debate on cartography proposed by Harley (2021), citing Foucault and Derrida, about the power and absences that maps incorporate takes on yet another status. When we bring together the assertions of Casé Angatu ('We don't own the land ... we are the land') and Denilson Baniwa ('There is no cartography in the world of the Pajés' [dos Anjos 2021]), it seems clear that a decolonial cartography is only possible by facing the crucial challenge of shifting from an anthropocentric to a ground-centred cosmology. If post-representational cartography warns us that the map is not the territory and the territory is not the map, then decolonial cartography is based on the assumption that another cartography will only be possible through other gestures of inhabiting the ground, when inhabiting the ground will be the same as cartography.

Thus, if there is a method for structuring the *Ground Atlas*, it intends to be open to and accept the infinite ways and methods of mapping. As an archive, it is basically made up of points and constellations. Points are single entries with very precise geographical coordinates, while constellations are configurations created from connections between selected points, with the aim of bringing together and mutually activating dispersed points to reveal new meanings and links between them.

In addition, it is possible to link points by layers—which offer other directions and reading possibilities—based on categories created by the *Atlas* editors, with the basic aim of classifying, cataloguing, and organising information and content. While the layers make up the strata from which the points are associated by similarities, the constellations are the rhizomes that enable transversal meanings.

As Figure 3 shows, each point is made up of hybrid and multiple components. In addition to geolocation, it has text that critically inscribes it in the context of historical disputes over the ground, with a set of iconographic references and tags or layers. Therefore, each point has a position but is always escaping from it. It is conceived as a critical node, an anchor for drawing constellations that make endless relationships and narratives visible and that cannot be separated from the beholder.

The search for plurality has also been a challenge in idiomatic terms. Despite the predominance, until now, of the language imposed through the European project of slavery and colonisation of what became Brazil, the *Ground Atlas* incorporates texts in original Amerindian languages, such as Guarani and Baniwa, and African languages, such as Kikongo, along with currently dominant languages, such as English and German. Likewise, following the call of Afro-Portuguese artist Grada Kilomba (2019), we pay attention to the relations of power and violence perpetuated by the Portuguese and give preference, for example, to the term 'enslaved', which describes a political project of dehumanisation, over the term 'slave', which suggests a supposedly natural identity. The dominance of the masculine gender in the Portuguese language has also received special attention in one of the most recent constellations (*Sara cura*) linked to the struggle for recognition of Black ancestry in a territory pressured by real estate speculation in the city centre of São Paulo[7].

From an iconographic point of view, we deal with the following conceptual and methodological questions: what status do, can, or should visual

7 Constellation linked to South Designs' Initiative *Designing with the Planet. Connecting Riverside Areas of Struggle in São Paulo, Jakarta and Berlin* (coordinated by Laura Kemmer and Jamie-Scott Baxter).

images have within the *Atlas*, considering the different formats and positions in which they are presented? What other meanings can emerge from the forms of articulation between the images present on the site, such as random and predefined montages, anti-narrative juxtapositions, and sequential and circular linearities?

From the point of view of the sources, hybridity is thus consciously explored; for example, photos taken by the cartographers coexist with images and films 'captured' from the internet, as well as sketches, audios, and Google Maps, while descriptive, factual, and fictional texts written especially for the *Atlas* coexist with texts taken from other sources. The aim is to compensate for the lack of a vertical dimension, which would somehow correspond to a topographic body, with a hybrid cartography that collects scraps and articulates multiple layers, sources, procedures, materials, categories, and temporalities in vanishing points.

Developed on the WordPress platform, the website also includes specific programming for random viewing and image movement, in addition to maps from Google chosen due to the possibilities they open up for interscalar displacement of the world map—intentionally centred between the African and American continents—from the localised experience of Street View.

An Independent Constellation

The most ambitious constellation so far in the *Ground Atlas* is related to the bicentenary of Brazilian independence, celebrated in 2022 by former Brazilian President Jair Bolsonaro's necropolitics. With the collaboration of dozens of researchers, artists, and activists from different backgrounds, nationalities, and disciplinary fields (architecture, history, geography, archaeology, visual arts, literature, and social sciences), the *Independent Constellation* comprises 200 points that evoke and problematise the processes of Brazilian independence and its developments from the standpoint of their relationship (historical and current) with the ground[8] (Figure 4).

This extensive mapping includes more obvious points, such as the Independence Monument in São Paulo (where the remains of the monarch D. Pedro, transferred from Portugal during the military regime, are buried) or the Valongo Pier in Rio de Janeiro (a landing platform for enslaved Africans whose recently unearthed ruins are now recognised as a UNESCO World Heritage Site). There are also less obvious points related, for example, to neoextractivism, such as Jacareacanga, one of the largest illegal mining areas in Brazil. In addition, there are points in other countries: in Portugal, a church in the city of Porto (where the heart of Pedro I is deposited and from whence it crossed the Atlantic Ocean in the most macabre celebration of the bicentenary of independence); in Benin, the city of Ganvie, which is believed to have been built over water to circumvent enslavement and consequently force displacement to what is now Brazil; in Athens, the temple where David Kopenawa saw the origins of the extractivism that is destroying Yanomami lands in the Amazon; and in Germany, Bayer's headquarters in Leverkusen, where pesticides now widely consumed in Brazil are produced despite being banned in Europe (Figure 5).

From this point of view, the *Atlas* plays a critical role in identifying, locating, bringing together, and making visible points linked to a long, tragic history of land disputes and conflicts in Brazil and accross the planet. This also means considering what it is to think about ground in Brazil—or from Brazil—today. In other words, the ***Atlas*** invites us to consider this history from a sociospatial position marked by colonial territorial logic. We stand at an unprecedented crossroads where the inequalities inherited from colonial violence are expressed as never before, in the very substrate on which we tread every day, in which we bury our dead, on which we build our houses, and on which the habitability of the planet depends. Notwithstanding the terrible effects of uncontrolled urbanisation, enclosures, extensive paving and sealing, predatory exploitation by mining and agribusiness, deforestation, policies of

8 The following cartographers took part in this constellation: Ailton Krenak, Ana Beatriz Queiroz, Ana Carolina de Paula Bezerra, Ana Clara Mesquita, Ana Luiza Nobre, Anita Ekman, Alexandre Silveira, Beatriz Campagnolo, Beatriz Gonçalves, Beatriz Tone, Beethoven Alvarez, Cândido Domingues, Carlos M. Teixeira, Carolina Pifano, Carolina Sacconi, Casé Angatu, Célia Tupinambá, Clara de Freitas, Clara Ianni, Coletivo Rio Tupinambá Karioka, Cristiana Lyrio Ximenes, Daniel Lavinas, David dos Santos da Conceição, David Sperling, Diego Inglez de Souza, Eleonora Aronis, Erica Lopo de Araújo, Erico Melo, Francisco dos Santos Ekman Simões, Frank I. Müeller, Gabriel Feltran, Gabriel T. Ramos, Giovanni Bussaglia, Guilherme Giufrida, Guilherme Wisnik, Heitor Martins Guimarães, Heloisa Gesteira, Joana Martins, João José Reis, João Masao Kamita, Julia França de Melo, Júlia Frenk, Julián Fuks, Juliana Alcântara, Laura Pappalardo, Ligia Nobre, Lisette Lagnado, Lucas Pereira Bosco, Luciano Bernardino da Costa, Maria Thereza Alves, Mariana Dourado, Mariza Bezerra, Marcelo Moscheta, Marcelo Motta, Maria Manuel Oliveira, Maria de Fátima Bento Ribeiro, Mariane Cardoso, Marina Gil, Moreno Pacheco, Paola Dargoni, Paulo Tavares, Philippe Braga, Piero Leirner, Priscila Faulhaber, Roberto Conduru, Ruy Sardinha Lopes, Silvana Andrade dos Santos, Thais Barcellos, Victor Vaccari Machado, Victoria Nassif, Wellington Cançado, and Wilmar D'Angelis.

erasure, banishment, deterritorialisation, soil contamination, erosion, and desertification—we can still hear the forest in the firm, yet tender voice of Ailton Krenak calling us to 'step gently' on the ground (Nobre and Calafate 2022, 139).

An Atlas as an Always-Unfinished Project

Just as every map is necessarily incomplete, every atlas is inconclusive by nature. So is the *Ground Atlas.* Yet, as a kind of spacetime mapping, it remains rooted in some principles: to write a non-linear and inconclusive story in which images take on the same status as words; to mobilise images—and relate words and images—to 'give dates a face' by considering the limits and potential of digital resources; to establish a dimension that has neither the verticality that hierarchises differences nor the horizontality that cancels them; to sustain the coherence of the ground as a composite of living and nonliving beings, elements, agents, dynamics, intensities, temperatures, states of composition, and decomposition; to foster within the dematerialised space of the internet the bodily understanding that underpins any knowledge of the ground; to set aside the dominant planetary vision and put the habitability of the Earth at the forefront; to explore the possibilities of operating critically with the Mercator projection that underpins Google Maps in order to reverse its Eurocentric and scientific logic in favour of other images and ways of positioning oneself in the world(s); to go beyond geopositioning according to two coordinates—longitude and latitude—which are certainly useful for precisely locating stable points but do not favour the mapping of interscalar dynamic processes and contested grounds; and finally, to give thickness to time in the stories that we aim to draw and redraw in the imaginative space of a digital atlas.

Image Credits
All Figures by Ana Luiza Nobre and David Sperling © *Ground Atlas*

References

Alves, Maria T. 2022. *Seeds of Change: New York, A Botany of Colonization, 2017–2018*. New York: Vera List Center for Art and Politics/Amherst: Amherst College Press.

Benjamin, Walter. 2007. *Passagens*. Belo Horizonte: UFMG.

Bispo dos Santos, Antônio. 2023. *A terra dá, a terra quer.* São Paulo: Ubu/ Piseagrama.

Deleuze, Gilles, and Félix Guattari. 1995. *Mil Platôs. Capitalismo e Esquizofrenia*. São Paulo: Editora 34.

Didi-Huberman, Georges. 2010. *Atlas ¿Cómo Llevar el Mundo a Cuestas?* Madrid: Museo Nacional Centro de Arte Reina Sofia; Tf Editores.

dos Anjos, Moacir. 2021. 'Para Decolonizar a Brasiliana'. *Select*, 5 February 2021. https://www.select.art.br/para-decolonizar-a-brasiliana/

Ferdinand, Malcom. 2022. *Uma Ecologia Decolonial: Pensar a Partir do Mundo Caribenho.* São Paulo: Ubu.

Guarani, Jerá. 'Para nós, o futuro é indígena'. 2022. In *Habitar o Antropoceno*, edited by Gabriela Moulin et al., pp. 8–21. Belo Horizonte: BDMG Cultural/Cosmópolis.

Haraway, Donna. 2016. *Staying with the Trouble: Making Kin in the Chthulucene*. Durham: Duke University Press.

Harley, John Brian. 2021. *The New Nature of Maps: Essays in the History of Cartography.* London: Johns Hopkins University Press.

Kilomba, Grada. 2019. *Memórias da Plantação.* Rio de Janeiro: Cobogó.

Kopenawa, Davi, and Bruce Albert. 2010. *A Queda do Céu: Palavras de um Xamã Yanomami*. São Paulo: Companhia das Letras.

Krenak, Ailton. 2020. *Ideias para Adiar o Fim do Mundo*. São Paulo: Companhia das Letras.

Latour, Bruno. 2018. *Down to Earth. Politics in the New Climatic Regime*. Cambridge: Polity Press.

Martins, Leda Maria. 2021. *Performances do Tempo Espiralar: Poéticas do Corpo-Tela.* Rio de Janeiro: Cobogó.

Mbembe, Achille. 2023. *La communauté Terrestre*. Paris: La Decouverte.

Nobre, Ana Luiza, and Caio Calafate, eds. 2022. *Sentidos do Chão*. Rio de Janeiro: Comum Pesquisas e Produções. https://www.atlasdochao.org/wp-content/uploads/2021/10/sentidos-do-chao_ebook_final.pdf

Puig de la Bellacasa, Maria. 2015. 'Making Time for Soil: Technoscientific Futurity and the Pace of Care'. *Social Studies of Science* 45(5): pp. 691–716.

Quijano, Aníbal. 2005. 'Colonialidade do Poder, Eurocentrismo e América Latina'. In *A Colonialidade do Saber: Eurocentrismo e Ciências Sociais. Perspectivas Latino-Americanas*, pp. 117–142. Buenos Aires: Clacso.

Segato, Rita. 2022. *Cenas de um Pensamento Incômodo.* Rio de Janeiro: Bazar do Tempo.

Shiva, Vandana. 2015. *Soil Not Oil: Environmental Justice in an Age of Climate Crisis*. Berkeley: North Atlantic Books.

Sperling, David M. 2024, 'Cartografias e Geopoéticas. Grafias e Poéticas de Mundos-Mais-que-Humanos'. *Questões Transversais - Revista de Epistemologias da Comunicação* 1: pp. 1–14.

Svampa, Maristella. 2019. *As Fronteiras do Neoextrativismo na América Latina.* São Paulo: Elefante.

Tavares, Paulo. 2020. *Memória da Terra: Arqueologias da Ancestralidade e da Despossessão do Povo Xavante de Marãiwatsédé*. Brasilia: Ministério Público Federal.

Tsing, Anna L. 2015. *The Mushroom at the End of the World*. Princeton: Princeton University Press.

Viveiros de Castro, Eduardo. 2013. *A Inconstância da Alma Selvagem*. São Paulo: Cosac Naify.

Warburg, Aby. 2010. *Atlas Mnemosyne*. Madrid: Ediciones Akal.

Jamie-Scott Baxter,
Vivien Sommer

Diffractive Mapping

Drawing Concepts Together/Apart in Spacetime Analysis

Introduction: Spacetimes in Social Change Research

This article outlines a specific spacetime mapping method that investigates socio-spatial phenomena through the ontological turn. This method is illustrated through its application in the field of research on social change, specifically social innovation. By employing Barad's (2003, 2007) diffractive method, the article focuses on the generative potential of drawing, reinterpreted through qualitative analytical practices rooted in Glaser and Strauss's (2008) grounded theory. This integration establishes a creative and analytical method we call *diffractive mapping*, tailored to address the specific research problem and explore socio-spatial phenomena in a broader sense. Diffractive mapping analyses the dimensions of spacetime in our study of social innovation across large urban-rural landscapes. Social innovation, seen as the intentional reconfiguration of social practices to address unmet social needs and foster new social relations, does not unfold in linear trajectories or static con-texts (Baxter 2023b, 2023c). Instead, it manifests as a performative process in which evolving practices and unfolding events continuously reshape spaces. This realisation led us to methodological challenges, such as capturing and interpreting practices that transform the very spatiotemporal structures they inhabit—challenges we address through our diffractive mapping approach and present in this article.

According to Barad (2007), practices are both material and symbolic, as they are in part carried out in particular locations, while also comprising discourses regulating, disciplining, and legitimising their materialisation. Therefore, we required a mapping method that would attend to the *material-discursive* constitution of practices (Barad 2007), locating them spatially and asking how they transform and perform transformation as they move and adapt with/in particular temporal events. In what follows, we describe the methodological challenges we faced.

Summary of Methodological Challenges

We required a programme that would help us consider the material-discursive and spatial implications of social change (social innovation) by researching two different cases of the transformation of large urban-rural landscapes. Therefore, the programme needed to be able to do the following: 1) map the historicity of events within each case; 2) reveal the practices and dynamics at work within and between specific events; 3) examine how practices intersect, connect, and adapt, thereby forming larger practice assemblages; 4) consider how discursive practices materialise and contribute to growth; 5) provide a visual and spatial description of the cases; register the performative qualities of social innovation that were becoming evident both theoretically and empirically; 6) investigate the ways in which temporal rhythms, sequences, and disruptions intersect with spatial figurations to shape innovation processes; and 7) explore how the relational dynamics of space and time inform the transformation and adaptation of practices across events.

Considering the dynamics of growth, change, and movement is a sociospatial challenge, one that we felt required a corresponding form of description and analysis. Together, we, the authors—a sociologist and an architect—began our endeavour by (re)thinking mapping and interpretation, traversing the boundaries of each other's disciplines in a diffractive mode. We attempted to understand in detail the specific analytical and generative components at work in each other's research practices.

In the following three sections, we describe what we found regarding the differences and entanglements between our approaches to reveal how design and drawing practices share an unexpected analytical and creative kinship with practices of coding and interpretation in grounded theory. In the first section, we introduce Barad's (2007) diffractive method and put it to work by reading *mapping*, *grounded theory*, and its extension, *situational analysis*, through each other. Here, we elucidate the differences and commonalties between these research programmes and suggest how each can compensate

Onto-epistemological elements	Map making	Grounded theory	Situational analysis	Diffractive mapping
Ontological				
Theoretical framework	Positivism	Symbolic interactionism	Social constructivism	Agential realism/ more-than-human
Position of reality (to data)	Direct (objective)	Mediated	Constructed	Co-generated
Epistemological				
Foci	Objects in/against space	Social phenomena	Social phenomena	Sociospatial assemblages
Data mode	Quantitative	Qualitative	Qualitative	Qualitative
Interpretative mode	Descriptive	Analytical	Analytical	Analytical
Interpretative protocol	Drawing	Coding	Coding/mental maps	Code-mapping
Reflexive mode	Normative	Reflective	Reflective	Agential
Communicative mode	Image	Text	Image/text	Image/text
Spacetime				
Spatial mode	Material	Social/relational	Analytical	Socio-material
Location of spatial concepts in analysis	Foreground	Background/context	Background/ metaphorical	Foreground
Temporal mode	Linear	Linear	Linear	Performative
Location of temporal concepts	Background	Background	Background	Foreground

Table 1 Diffractively reading grounded theory through mapping (Jamie-Scott Baxter and Vivien Sommer, 2020)

for deficiencies in the others. We also outline our proposed methodological intervention, which we have termed *diffractive mapping*. In the second section, we describe the three modes of our diffractive mapping method in detail. For the sake of clarity, we present these as distinct, but in actuality, the boundary between modes is nebulous and permeable, as one flows into the next and circles back to the others. Each mode is illustrated with a drawing taken from our case studies about the micro practice of social change in large landscapes in Portugal and Austria (see Baxter 2023a, 2023b, 2023c). Finally, the paper ends with a summary of our argument and a reflection on how our diffractive mapping methods meet the requirements set out in our research design.

Diffractive Method

We refer to our approach as *diffractive mapping*, borrowed from the diffractive method that suffuses Barad's agential realist theoretical (2003, 2007) intervention into the philosophy of science. Building on feminist scholars Donna Haraway and Trinh T. Minh-ha, Barad elaborates 'diffraction as a tool for analysis attending to and responding to the effects of difference' (2007, 72). Haraway (1992, as cited in Barad 2003, 803) tells us that 'diffraction is a mapping of interferences, not of replication, reflection, or reproduction'. Barad goes on to argue that their use of diffraction is 'a mutated critical tool for analysis'—that is, for mapping the *effects* of difference (Barad 2007, 72). In this way, diffraction refers to traceable marks left on bodies as material-discursive practices intersect and intra-act. Diffraction challenges dualities such as 'Nature' and 'Culture' by demonstrating that externality exists only within other practices (Barad 2003, 803). Diffraction is both an epistemological and ontological consideration, as research apparatuses not only observe but also actively shape the research object and its materialisation (Barad 2007, 161). Drawing on this, we use the concept to describe how mapping and coding practices intersect, particularly as mapping techniques are integrated with grounded theory protocols, and how research apparatuses themselves shape the spatiotemporal dimensions of socio-spatial phenomena.

Mapping Practices

Maps, like drawings, are powerful generative apparatuses that not only represent spatial relations but actively co-constitute spatialities (Crampton 2009; Kitchin and Dodge 2007). Mapping, like drawing, depends on the generative practice of making marks, a process described by Evans (1986) (diffractively read through Barad) as an intra-action in which the outcome is not fully predictable, emphasising the agential nature of drawing and mapping in producing new possibilities. Maps simultaneously describe and create spatial relations, leaving traces that materialise phenomena over time.

Although maps are often seen as objective representations (Harley 1987), critical literature highlights their embedded power dynamics and political nature, with maps acting as discursive practices with material consequences (Pickles 2004). Mapping is a material-discursive and performative practice but is also influenced by normative assumptions and stylistic biases, particularly in spatial design disciplines for which maps are tools for planning rather than robust methods for spatial analysis (Baxter et al. 2025). Recognising diagrams as maps reinforces this argument since both diagrams and maps function as tools for visualising and co-producing spatiotemporal relations, thus demonstrating their potential to serve as methodologically rigorous tools for understanding and shaping the intertwined dynamics of space and time.

Grounded Theory and Situational Mapping

Grounded theory aims to develop theories grounded in data through reflexive, comparative analysis (Glaser and Strauss 2008). It balances openness with rule guidance, with rules serving as flexible guidelines (Strauss 2003). Its circular analytical process involves three steps of coding—open, axial, and selective—that move from concrete analysis to abstract conceptualisation, while continuously revisiting earlier stages. Coding enables theoretical concept development through constant data comparison rather than mere empirical description (Strauss 2003). Mapping is integrated into coding, particularly in axial and selective stages, where codes are related through matrices, although spatiality and temporality often serve only as background contexts (Strauss and Corbin 1990).

Clarke's (2005) situational analysis expands grounded theory to address complexity, emphasising that research constructs reality rather than revealing it. Situational analysis uses maps to visualise interpretation, including situational maps (showing relationships between human, nonhuman, and discursive elements), social world/arena maps (focusing on organisational and discursive dimensions), and positional maps (highlighting discursive positions) (Clarke 2005; Clarke et al. 2015). However, spatial and temporal dimensions remain secondary. While Clarke's mapping tools reflect a postmodern turn, they do not fully harness the generative potential of maps nor do they prioritise the explicit analysis of spacetimes or the process of transformation, thus limiting their capacity for analysing practices that produce (and are shaped by) spatiotemporal dynamics.

Ethico-Onto-Epistemology and How Spacetime Matters

We acknowledge that methods of map making, grounded theory, and their evolutions are based on different theoretical assumptions regarding *reality* and its relationship to data, which are themselves often caught up in questions of access, representation, mediation, and (re)construction. Classical, or traditional, cartographic mapping procedures, for example, lean towards a positivist ontological position, in which the world is measured and a representation is made to present facts. Such positivist foundations have, of course, been contested in the critical cartography literature, as we have briefly discussed. This critical approach is closely aligned with the symbolic interactionism underlying grounded theory. These assumptions influence how data and the function of data analysis are defined. In grounded theory, analysis is based on an indicator-concept model, in which concepts discovered in data are an indicator of reality. Through this framework, researchers 'discover' concepts by comparing indicators over and over again (Strauss 2003, 25). Clarke's (2005) situational analysis contests this idea and argues that concepts are not 'discovered' in the data but are 'constructed' through research. In Barad's (2007) agential realism, realities are generated through material-discursive practices constituted, at least in part, by more-than-human forces, through which the possibility of agency is determined not *a priori* but within these intra-actions. This ontological position allows matter to play an agential role in the process of materialisation, thereby accounting for the co-generative forces between human and nonhuman bodies. We suggest that this ontology is captured and performed through the generative practices of drawing and coding in our diffractive apparatus, an apparatus that extends beyond the research programme.

To meet our research objectives, our search for a methodological programme rests on the assumption that research is performative and consequential—that is, that research practices play a part in the materialisation of the object of investigation. Indeed, in Barad's (2007, 185) *ethico-onto-epistemology* there is no prior object-subject division; rather, all boundaries, cuts, and exclusions are performed through the intra-active meeting of agencies in practices (Barad 2007). It is not the static image we strive for in our *code mapping*, but rather the generative characteristics of drawing as an intra-active material-discursive apparatus that leaves marks on bodies which can be retraced and are accountable.

Diffractively and together, we, the authors, can address some of the other binary distinctions—such as those performed through disciplines, cultural backgrounds, and gender—but not all. Tools in grounded theory can help reflect upon the normative judgements at work in mapping. Coding data is often a hierarchical and linear procedure, but by using drawing to generate concepts, the relations between data and codes are emphasised and visualised. As we illustrate in the following section, in our diffractive approach, codes are 'drawn out' of the data and connected cartographically, allowing for multiple connections and relations to be made as concepts are co-generated and mapped. This works to undermine contrastive thinking, in which boundaries are drawn around concepts and exclusions made. However, this form of thinking can easily lead to unhelpful dualities and categories.

A significant requirement of our research programme is to understand how space and time are performed through practices of social innovation —that is, how practices are not only produced in events but also co-constitute them. We aim to foreground the concept of space and to consider its material and relational presence—that is, how spatialities are generated within intra-action, but also how they are agential within this intra-activity.

To summarise this discussion, or rather to articulate it in another way, we present Table 1 highlighting the commonalities, divergences, and assumptions between and within these approaches, as identified above.

Doing Diffractive Mapping

Our approach moves between three spatial and analytical modes. The *first mode* provides a new base layer consisting of the events and processes at work in the cases. The *second mode* plots, through these events, the specific practices relevant to the research question; in our case, it identifies and locates which practices were on the move in the case and how they intra-acted with other practices/entities. Initially, these rich and messy arrangements of practices and events are rendered in their complexity following a heuristic process of gradual abstraction and reiterative circling back until a conceptual diagram emerges in the *third mode,* which describes the contours and concepts of the socio-spatial arrangement.

We would like to emphasise that our approach requires no fixed number of drawings (e.g., there could be one drawing with three layers following the modes, or three or more separate drawings), nor is there a specific style of drawing. Rather, the heart of our cartographic approach is the process of building up through the modes and circling back through them reflexively until saturation is reached in order to co-generate a visual and spatial description. Furthermore, a benefit of this approach (but certainly not uncontested, cf. Latour 2018) is the ability to effortlessly 'zoom in' to spaces within the map to understand detail and, conversely, 'zoom out' to get an overview or connect different events. This is reminiscent of Smith's (1992) *scale jumping*. In our case, this was done digitally using basic drawing software.

Mode 1: Base Plan—Events and Processes

A base plan in typical map making and architectural drawing is usually a flat, one-dimensional visual description of a territory, such as an urban neighbourhood, a geographic region, or part of an existing building. It describes a *given* spatial arrangement, usually prior to a mapping or design task being carried out, because there needs to be some form of contrast in order to map something. Usually, this is done through the rendering of 'nature' (through geographic representation) as static to provide a stable backdrop against which the 'liveliness' of social activity and its agency can be mapped. However, as we have discussed, nature can no longer be performed as an inert and stable background where the social unfolds, unmarked by the 'context' upon which it acts. So, what of our base layer? Similar to Strauss and Corbin's (1990) open coding, which asks basic questions pertaining to who, when, where, what, how, how much, and why, our first mapping mode was

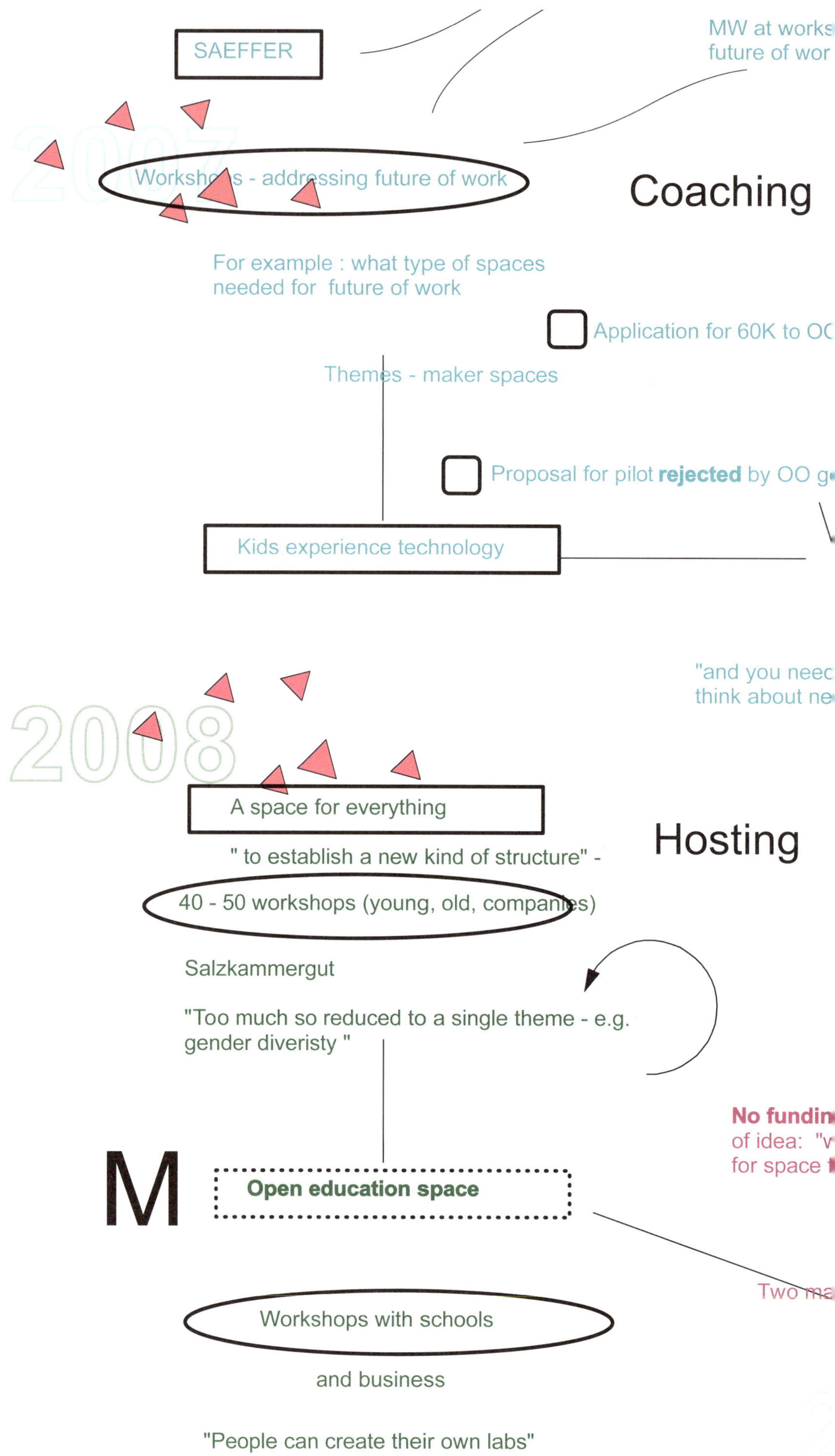

Figure 1 Excerpt from base plan: events and processes (Jamie-Scott Baxter, 2020)

lked alot about the

Travel and visting

: project ☐

nt ☐

"Young people leaving the countryside"

g space to
for innovation"

Figure 1. Population change between 2000 and 2014 by NUTS 3 regions

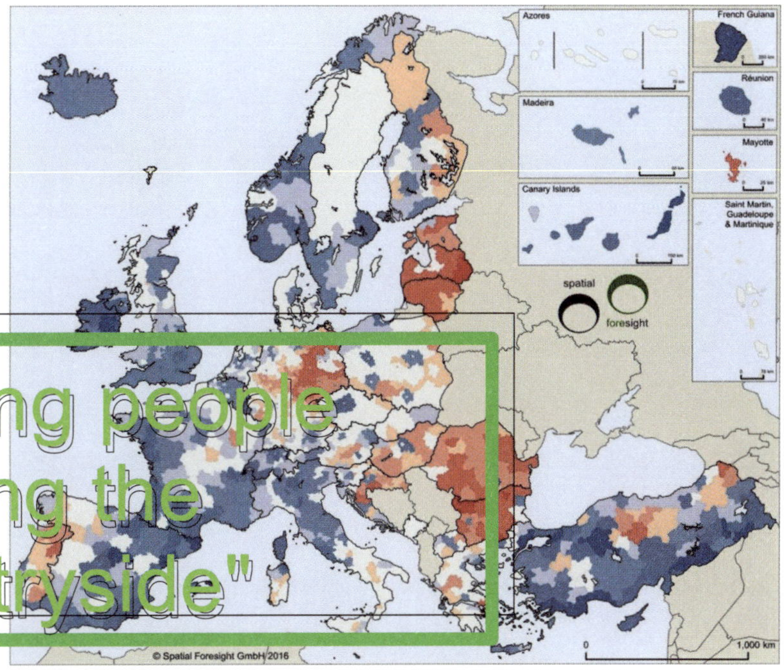

Administrative boundaries: Spatial Foresight and University of Geneva based on material from Eurostat GISCO, the GADM database and the EEA.
Data: Eurostat (demo_r_pjangrp3).
Reproduction permitted for non-commercial purposes with full referencing of Spatial Foresight.

Annual average population change from 2000 to 2014 in %

-3 - -2
-2 - -1
-1 - -0.5
-0.5 - -0.25
-0.25 - 0.25
0.25 - 0.5
0.5 - 1
1 - 2
2 - 4.5

Data for AT, IT, LV, HR, HU and MT from 2001 - 2014.
Data for SK, ES, CZ, UKD6, UKD61, UKD62, UKD63, UKD71, UKE44, UKE45, UKF24, UKF25, UKG36, UKG37, UKG38, UKG39, UKH24 and UKH25 from 2002 - 2014.
Data for NL, SI and IS from 2003 - 2014.
Data for BE335 and BE336 from 2004 - 2014.
Data for NO011 and NO012 from 2005 - 2014.
Data for DK from 2007 - 2014.
Data for TR from 2008 - 2014.
Missing data for parts of DED2, DED4, DED5 and DEE0 was subsituted with data from the next higher NUTS level.

Source: Spatial Foresight.

2008 financial crash followed by recession

"Traditional technolpgy focused"

ificant to development
d to ask muncipalites

Funding available

"Only chance that a municipality /local government will fund it"

"yes"

First open labs established

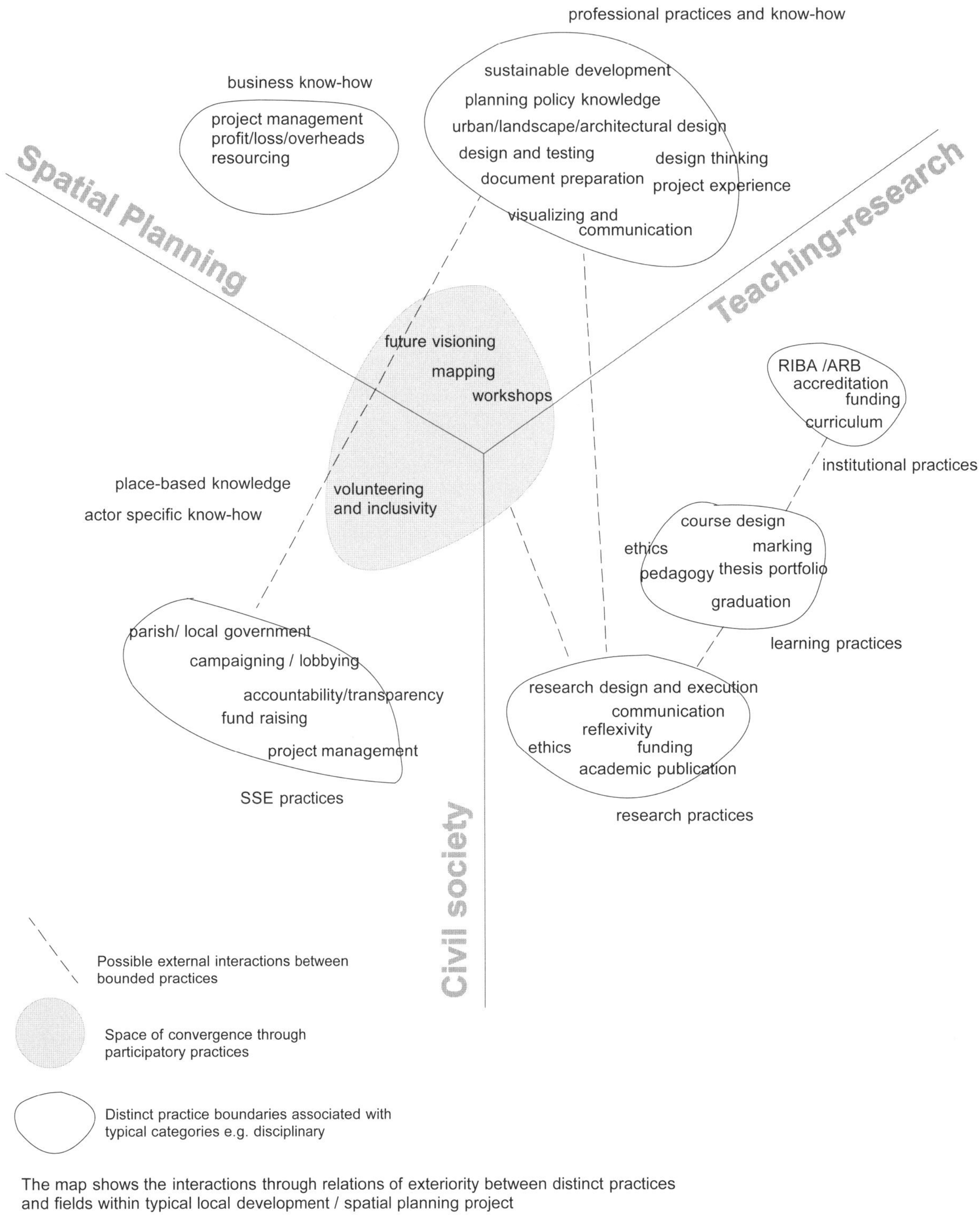

Figure 2 Figure-ground and practices (Jamie-Scott Baxter, 2020)

guided by the following question: *which events have been significant in the history of the specific social phenomena under investigation?* Events and processes are both temporal and spatial and are unlikely to be misunderstood as static. Nonetheless, this layer provides a (lively) 'context' derived from the data, against which further phenomena are then mapped. In our case, and in the example below, the research objects were 'open innovation labs' in Austria (Figure 1).

The history of the emergence of these labs is relatively short: the first lab opened in 2011. We plotted events back to around 1996, which were relative to the emergence of the first and subsequent evolutions of the labs as the assembly grew. The events were captured through over thirty semi-structured interviews with participants who were either directly involved with the co-generation of new labs in new locations or were public servants (administrative or elected) and would have been aware of the story (or may have affected its course). In addition, we also managed to speak to participants who were not directly involved with the evolution of the labs. These, at times, converging stories of events given by multiple participants were mapped together to form a dense, 'contextual' surface—or *base plan*—for our emerging map. Where we required further granularity, we carried out additional document analysis, which was referenced in the map. As in open coding, in which data are initially 'cut out' from the empirical case, this mode provided the first level of abstraction and allowed us to compare similarities and differences within the data and to look for patterns as connecting practices started to emerge.

Mode 2: Figure-Ground—Practices and Intra-Action

The second mode maps the practices and intra-actions at work within and co-constituted by the events and unfolding processes of the first layer. Here, the assumption, which follows in the footsteps of a rich history of practice theory (see, e.g., Reckwitz 2002; Schatzki 1996; Shove et al. 2012), is that practices are the location of the social. Indeed, according to scholars working through the so-called spatial turn (see, e.g., Harvey 1996; Hillier 2007; Lefebvre 1991; Löw 2008; Massey 2005; Thrift 1996), it is in action that space, or spatiality, is performed. Hence, in socio-spatial research, we attend to *the events, processes, practices,* and *their relations*. In our case, the objective was to shed light on the practices at work within a particular social innovation constellation and the dynamics that propelled its growth. We mapped the practices that connected events and began to codify them. Much like in the axial coding of grounded theory, our focus was to provide a deeper interpretation of the data and consider the relationships between emerging practice codes and how they intra-act without nesting them into categories.

The second mode is thus guided by the following question: *what specific (material-discursive) practices suffuse the events in the base plan, and how do they intra-act?* In our case, the question was refined to focus on how social innovation constellations grow as particular material-discursive practices connect and adapt to each other within different events, where events not only *take place* (locatable) in a time but also *affect* time and spatialities through practices (Figure 2).

Figure-Ground Practices

Through our mapping analysis, we were able to plot the practices and relations as new emergent hybrid practices were (per)formed. We were able to see how certain discursive practices (e.g., brand names and logos) were being used to connect events across territories and how practices that existed in specific locations (e.g., radio stations and broadcasting) adapted to practices that were more mobile (e.g., hosting and training-the-trainer), claiming new forms of open innovation labs. This led us to understand that there was no social innovation model being acted out, no *best practice* being simply repeated, but rather new practices performed 'for another first time' (Garfinkel 1967) as practice bundles connected, merged, and adapted. This meant novelty and

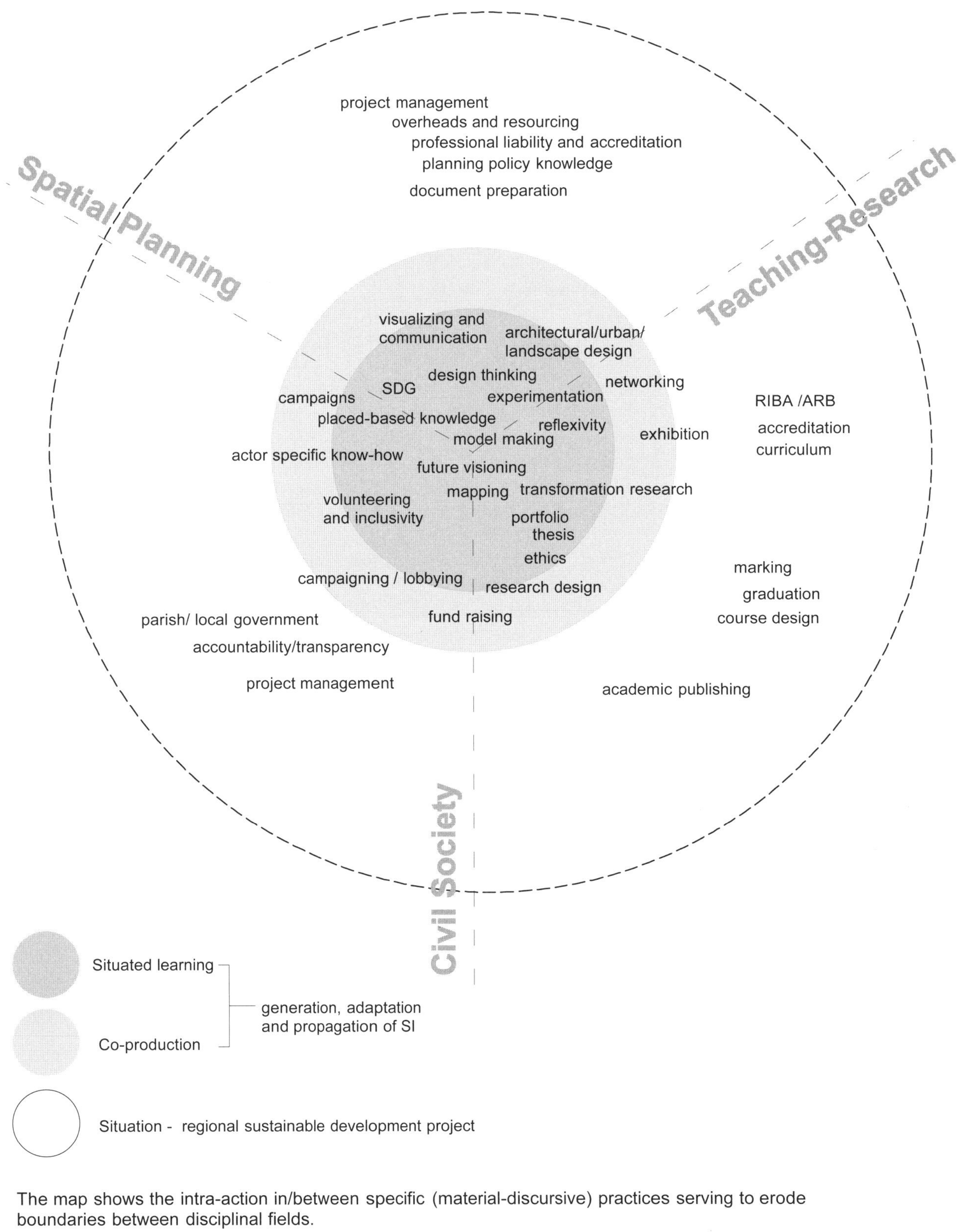

Figure 3 Contours and concepts (Jamie-Scott Baxter, 2020)

growth occurred simultaneously. Through our mapping protocol, we developed the concepts of *intrinsic practices* and *practices exterior-within* to capture those practices that existed in an event (e.g., in a location or in an organisation) and those that had been transferred, at least within the time frame of our research.

In architectural drawing, a figure-ground representation is constructed through the visual contrast of two layers. The stark contrast of a black figure against a white 'background' results in a spatial and reciprocal relationship between the two parts, with the 'background' taking on its own figurative quality through the contours of the 'foregrounded' shape. This tool is often used to describe the form of space between objects, with object and space read in relation to and though one another in a single flattened map. We borrow this term to highlight the intra-action between layers in which the background and foreground are reciprocally performed through each other—in our case, between practices and the event through which they arise and are co-constructed.

Mode 3: Socio-Spatial Diagram—Concepts and Contours

The third mapping mode is the culmination of a heuristic process of circling back through the previous layers towards an abstract visual diagram of the empirical case until a point of saturation is reached. As in the selective coding of grounded theory, in which the aim is to formulate a central narrative that provides an overview of the data and their connections, the questions that guide this mapping mode are as follows: *which new concepts arise from the case, and what is the shape, size, and density of the new socio-spatial arrangement at large?* For example, in Figure 3, we see new hybrid categories of *joint problematisation* and *joint learning*, the latter consisting of social, situated, and mutual learning practices that serve to propagate social innovation in our case. Like concepts, these practices span other categories, which, in this example, correspond to the fields of *spatial planning*, *civic society*, and *academia*. What is evident in the images is that they are not branches of a tree-like knowledge structure but connect and span rhizomatically across and through one another. Similarly, the diagrams show how the boundaries, or contours, between these fields are permeated by other constituent practices. This form of mapping allows us to register more complex relations between concepts and the reality from which they emerge and perform (Figure 3).

Socio-Spatial Diagram: New Contours and Concepts

As in grounded theory, in which concepts aim towards more generalisable theoretical formations that can be compared across cases, the diagram can be used to analyse, compare against other diagrams from other cases, and aid in communicating results. Perhaps more urgently, the diagram, like all practices of drawing, has the ability to communicate back to its makers. The reflexivity built into this mode is paramount to our mapping protocol and was complemented by memo-taking throughout the programme and as we mapped. By making memos, we began to question instinctive actions in map making, especially those surrounding aesthetic decisions, which—beyond their visual impact—would often go unquestioned during conventional map making. These reflexive gestures address the deficiencies in spatial planning and architectural map making highlighted in Table 1.

We set out to provide an analytical apparatus that would fulfil the following criteria: map the history of events within a specific case; reveal the practices and dynamics at work within and between these events; examine how practices intersect, connect, and adapt to form larger assemblages; consider how discursive practices materialise and contribute to growth; provide a visual and spatial description of the cases, and register the performative qualities of the social and spatial.

We met this challenge by combining insights from grounded theory with practices of map making in order to fill, on the one hand, the methodological gap of material and relational spacetime analysis in qualitative research and, on the other, the missing reflection about data interpretation and analysis in mapping practices of architecture and spatial planning. By bringing grounded theory together with map making, the emphasis was on spatial and material relationships and intra-activity at work in spacetime processes. However, while grounded theory looks for a hierarchy in coding to develop a key category, we used a cartographic approach. Instead we mapped the genealogies of practices which suffuse specific events, making connections through and across events in order to look at how, for example, transformative practices adapted and evolved and how they affected the spatialities they permeated as they travelled. By mapping the relations and intra-actions between events and practices, we provided a visual, spatial, and performative method restrained by reflexivity which describes the world it co-generates.

We argue that focusing on events, processes, practices, and their relationships—instead of on actors and their (human) networks—in social change research requires not only a new set of tools but also an ontological shift. We searched for the diffraction patterns left by practices as they propagate, wave-like, through events, while co-constituting the processual medium they suffuse. Such patterns are the result of intersections and adaptions as they move through each other and around obstacles in their paths. These diffractive marks and patterns are what our diffractive mapping method records; in our case, these were the practices associated with social innovation that propagate through large urban–rural landscapes. As we have discussed, space is not an inert medium through which practices travel unscathed. Rather, specific spatialities are co-constituted by events, processes, and practices and by the relationships between them. Aligned with the ontological turn in social theory, which decentres human agency, our diffractive mapping and analysis provides a nuanced way to attend to the entangled material and social dimensions of spatiotemporal phenomena.

Image Credits
Table 1 by Jamie-Scott Baxter and Vivien Sommer
Figures 1, 2, 3 by Jamie-Scott Baxter

References

Austin, John. L. 1962. *How to Do Things with Words*. Oxford: Oxford University Press.
Barad, Karen. 2003. 'Posthumanist Performativity: Toward an Understanding of How Matter Comes to Matter'. *Signs: Journal of Women in Culture and Society* 28(3): pp. 801–831.
Barad, Karen. 2007. *Meeting the Universe Halfway: Quantum Physics and the Entanglement of Matter and Meaning*. Durham: Duke University Press.
Baxter, Jamie-Scott. 2023a. 'Dis/Continuing In/Justices-to-Come in "Ghost Town" Nava Raipur India'. *lo Squaderno* 64: pp. 15–19.
Baxter, Jamie-Scott. 2023b. 'Modes of Spread in Social Innovation: A Social Topology Case in Rural Portugal'. *Journal of Rural Studies* 99: pp. 243–251.
Baxter, Jamie-Scott. 2023c. *Self-Organised/ing Infrastructures. Understanding Everyday Practices of Spread and Resistance in Processes of Socio-Spatial Change on the Periphery. Multisited Ethnographies in Rural Portugal and Austria.* PhD thesis, Technische Universität Berlin.
Baxter, Jamie-Scott, Séverine Marguin, Sophie Mélix, Martin Schinagl, Ajit Singh and Vivien Sommer. 2025. 'For Hybrid Mapping: A Methodology to Study Relational Space. The Case of Botanical Gardens Berlin'. *Architecture and Culture* 11(3).
Clarke, Adele. E. 2005. *Situational Analysis: Grounded Theory after the Postmodern Turn*. Thousand Oaks: Sage.
Clarke, Adele E., Rachel Washburn, and Carrie Friese, eds. 2015. *Situational Analysis in Practice: Mapping Research with Grounded Theory.* Walnut Creek: Left Coast Press.
Crampton, Jason. W. 2009. 'Cartography: Performative, Participatory, Political'. *Progress in Human Geography* 33(6): pp. 840–848.
Evans, Robin. 1986. 'Translations from Drawing to Building. AA Files'. *Historical Research* 40(3): pp. 258–278.
Garfinkel, Harold. 1967. *Studies in Ethnomethodology.* New Jersey: Prentice-Hall Inc.
Glaser, Barney. G., and Anselm. L. Strauss. 2008. *The Discovery of Grounded Theory. Strategies for Qualitative Research.* New Brunswick: Aldine.
Harvey, David. 1996. *Justice, Nature, and the Geography of Difference.* Cambridge: Blackwell Publishers.
Harley, John, B. 1987. 'The Map and the Development of the History of Cartography'. In *The History of Cartography*, edited by J. B. Harley and D. Woodward, pp. 1–42. Chicago: University of Chicago Press.
Hillier, Jean. 2007. *Stretching Beyond the Horizon: A Multiplanar Theory of Spatial Planning and Governance*. London: Routledge.
Kitchin, Robert., and Martin. Dodge. 2007. 'Rethinking Maps'. *Progress in Human Geography* 31(3): pp. 331–344.
Latour, Bruno. 2018. *Down to Earth: Politics in the Climatic Regime* (C. Porter, Trans.). London: Polity Press.
Lefebvre, Henri. 1991. *The Production of Space.* (D. Nicholson-Smith, Trans.). Cambridge: Blackwell.
Löw, Martina. 2008. 'The Constitution of Space'. *European Journal of Social Theory* 11(1): pp. 25–49.
Massey, Doreen. 2005. *For Space*. London: Sage.
Pickles, John. 2004. *A History of Spaces: Cartographic Reason, Mapping, and the Geo-coded World.* Abingdon, UK, and New York: Routledge.
Reckwitz, Andreas, 2002. 'Toward a Theory of Social Practices. A Development in Culturalist Theorizing'. *European Journal of Social Theory* 5(2): pp. 245–265.
Schatzki, Theodore, R. 1996. *Social Practices: A Wittgensteinian Approach to Human Activity and the Social.* New York: Cambridge University Press.
Shove, Elizabeth, Matt Watson, and Mika Pantzar. 2012. *The Dynamics of Social Practice: Everyday life and How it Changes.* London: Sage.
Smith, Neil. 1992. 'Geography, Difference, and the Politics of Scale'. In *Postmodernism and the Social Sciences*, edited by J. Doherty, E. Graham, and M. Malek, pp. 57–79. London: Macmillan.
Strauss, Anslem, L. 2003. *Qualitative Analysis for Social Scientists*. Cambridge: Cambridge University Press.
Strauss, Anslem and Juliet Corbin. 1990. *Basic of Grounded Theory Methods*. Beverly Hills: Sage.
Thrift, Nigel. (1996). *Spatial formations.* SAGE Publications Ltd. https://doi.org/10.4135/9781446222362.

Jae-Young Lee

Experiential Mapping

Urban Experiences in a Mountainous Rural Place in South Korea

Learning 'the Field' as a Research Tourist

In recent years, scientific discourses have increasingly deconstructed rural-urban binaries, shifting towards more processual and territorial understandings (Robinson et al. 2022; Schmid and Topalovic 2023; Sun 2023). Large-scale development projects have been interpreted within frameworks of 'variegations' of wider circulating processes, such as neoliberalisation or financialisation. However, such projects often entail significant state support and investment, are strongly linked to a wide variety of transnational investors and developers and are frequently highly contested in their local environments. Thus, each project comes to fruition in a distinctive regulatory context, often as an exception to the norm, and each emerges through complex interactions over a long period of time among an array of actors. We therefore seek to broaden the discussion from an analytical focus on variegated globalised processes to consider three large-scale urban development projects (in Shanghai, Johannesburg and London. Nevertheless, rural–urban binaries remain pervasive heuristics in contemporary spatial imaginaries and planning. This is corroborated by empirical findings that emphasise situated spatial ontologies in rural places (Pavez et al. 2017) and possible trajectories of rural peripheries that may look 'quite different from what we see in the core' (Pugh and Dubois 2021, 274). Exploring the unequal power dynamics between emic and etic perspectives might lead to these discursive hegemonies, and urban scholars are thus highlighting the value of spatial experience. Simone describes urbanisation as a proliferation of 'territories of experience' that turn experience into 'a commodity, as something bearing particular symbolic values' through the 'colonialisation of the sensibilities and bodies of diverse people' (Simone 2023, 383) and certain spatial configurations. Aligning with this, Robinson (2022, 204) defines the researcher as a 'theorizing subject' who is very much entangled in such power dynamics.

This article takes the theorising subject of the researcher as an entry point for a reflexive methodological mapping exploration in which my own field access is objectified. Drawing on McFarlane's concept of *urban learning* (McFarlane 2011b) as a way to trace how subjects move through and make sense of space, this field access is conceptualised as my spatial learning process. However, while McFarlane focuses on *urban* environments, this auto-ethnographic exploration is set in a mountainous place in South Korea, considered *rural*. By following my own embodied experiences of getting to know a place, this methodology aims to trace and disentangle spatial experiences from rural/urban discursive attributions. It asks the following questions: what were my own perceptions and practices in space during fieldwork? Which actors and processes co-constituted these perceptions? How did my perceptions correlate with the residents' spatial perceptions and practices that I set out to study?

These observations took place within the subproject B01 *Peripheralised Rural Areas: Digitalisation and Constructions of Space* of the DFG-funded CRC 1265 *Re-Figuration of Spaces* at the Technische Universität Berlin, Germany. Initially born as a coping mechanism, my visual attempts to gain control over my own perceptions eventually inspired this article.

Mapping as a Gateway to Experiential Knowledge

In efforts to sensitise the sociomaterial dynamics that might constitute 'the urban', McFarlane (2011b, 175) theorised that, in order to move through and in urban space with agency, each subject undergoes an 'intersubjectively valuable learning' process. McFarlane's *urban learning* outlined a negotiation process between multiple agencies of multiple actors in space that, together, shape not only power dynamics but also their consolidation in spatial arrangements and dynamics. This field of negotiation between agencies was conceptualised as *urban assemblage* (McFarlane 2011a). Such materialist analytical lenses brought forth conceptualisations of *rural* assemblages (Lendvay 2021; Woods 2016) and more-than-human dimensions to the political ecology in rural areas (Brown et al. 2019; Gillon 2014). Nevertheless, in more political-economic approaches to nonurban areas (Arboleda 2016; Markaki 2023), the subject matter of the researcher's mind and body has been largely overlooked. This has manifested in the critique of 'disembodied theory' (Peake 2016, 3), shared by critical urban scholars calling for the inclusion of a broader range of perspectives and experiences.

In this respect, the methodology of mapping has caught multidisciplinary attention. Within a pedagogical geographical context, Elwood (2004, 62) posited that students' mapping of their own experiences as 'experiential practice' can visualise their own dispositions to reflect upon conceptualisations of the urban. Moreover, subjective sketch mapping has been investigated as a reflexive methodology for the analysis of socio-spatial perceptions of geographical space that has the potential to reveal 'the emergence of new interpositional "trends" and patterns and perceptions that influence the research design itself' (Hawthorne et al. 2015, 34). Such autoethnographic and experiential mappings are therefore credited with the ability to visualise and objectify one's own emplacement as multisensory entanglement within the more-than-human assemblages of a place (Powell 2016), thereby offering revelatory potential to access the 'elusive, hidden or underrepresented' (Powell 2010, 554) qualities of places in relation to our position. Langran and DeWitt (2020) framed this multimodal approach as 'critical place-based inquiry', situating it within the discourse of urban political ecology. They highlighted the potential for understanding the interconnections between 'the local, the regional, and the global contexts' (Langran and DeWitt 2020, 60). However, experiential mapping has primarily been explored as a reflexive teaching methodology to illustrate the significance of positionality.

The objective of this article is to extend this function by examining autoethnographic experiential mapping in correlation with the data collected from the residents of the above-mentioned mountainous place regarding their spatial perceptions and practices. By exploring the ephemeral gap between my own initial and later perceptions of this rural place, which was co-constructed by residents, this article aims to assess the potential of such mappings for new insights.

Case Study, Positionality, and Methodology

The fieldwork was conducted over eight consecutive weeks in a mountainous rural valley in South Korea between July and September 2023. For the purposes of this article and to protect the privacy of all parties engaged, the valley is henceforth referred to as Noram. Noram is a rural village entity made up of several settlements constituting around 2,500 inhabitants in 2023 (Korean Statistical Information Service 2023). The problem-centred and go-along interviews with residents engaged in platform tourism and e-commerce focused on their spatial perceptions of and practices in Noram. For the selection of the field study area, aside from a literature review on rural digitalisation, both my language and cultural proficiencies as a researcher played a significant role.

As a Korean-born, cisgender, female migrant-turned-citizen in Germany, I had experience living both in Seoul, South Korea, and in various remote, rural places in Germany. By the beginning of this case study in 2023, I was a trained architect, having been practising architecture with my colleagues at the rural Thai–Burma border for almost eight years. With this specific gaze on my own biography, I hoped for a mutual recognition from the rural residents of the fact that we were all coming from the 'periphery', a geographical state publicly established to be ever-deficient in urban places. At the same time, I had to look at my frameworks. My parents had brought our family to Germany for my father's doctoral studies when I was three years old. Through them, higher education in various cities had always been socially and economically accessible to me, leading to this very fieldwork, having struggled to attain an academic degree in the capital of Germany. This also meant that, all my life, I had been trying to digest and engage with discourses from a mostly White,

Eurocentric elite that were firmly fixated on the urban metropolises of the world.

To counteract my own biases, Seoul-based sociology student Sungwon Ryu was set to join the fieldwork as an assistant. However, we chose for her to arrive only after three weeks, during which I would first have to introduce myself to the villagers without the benefit of mediation. The first mappings thus started in this period. Leaning on the concept of *urban learning*, the daily mapping processes focused on my encounters with the more-than-human actors around me, how I made sense of them, and how my relation to them informed my perceptions and movement in space. This entailed not only daily drawings and trackings of my movements but also spontaneous and retrospective sketches, interviews with residents, and field notes.

The resulting datasets were operationalised in three multimodal *joint spatial displays* (Marguin et al. 2021). To facilitate the visualisations of my own spatial experiences, four categories were selected to be represented in the three maps: 1) daily movement by foot (movement by car was not documented), 2) areas perceived as hostile, 3) areas perceived as open to the public, and 4) areas perceived as privatised public spaces.

For more depth, three additional descriptive categories were introduced: 1) areas perceived as 'alien' (to the local fabric), 2) areas perceived as 'local' infrastructure, and 3) spatial codes as sketches of places and situations that caused spatial discomfort, underpinned by field notes.

The influence of residents' spatial perceptions and practices on my spatial learning was incorporated through direct quotes from interviews and through participatory observations. In terms of temporality, the maps encompassed three distinct phases: 1) monsoon season, 2) national vacation season, and 3) post-vacation summer season.

For the purpose of this article, the information density in the presented maps is limited to the stories and fragments outlined in the analysis.

From 'Hostile' Spaces to Subjectified Places

Choosing a mode of transport provided a reflective moment even before fieldwork had begun. My romanticised projection of a walkable European '15-minute city' in a rural Korean village was swiftly corrected by friends and family in South Korea, who advised me to rent a car. However, walking within the village was chosen as a strategy to gain field access and get to know the place. Here, my lack of spatial knowledge forced me to rely heavily on processing the spatial cues of built environments and on paying attention to the dynamics of more-than-human actors to ensure my bodily safety.

'Hostile' spaces, as the first established category, encompassed spaces of my own discomfort. These were initially characterised by signs of human abandonment and the strong presence of nonhuman actors, such as houses completely ingested by plants (Figure 1, Sketch 3). The steep altitude leading up to a village, accompanied by built structures, such as village gates (Figure 2, Sketch 5), emerged as an active threshold that deterred my exploration of certain areas in the beginning. Similarly, densifying humidity and heat within the mountain forest triggered outpourings of sweat, eventually convincing me to abandon my plan to visit a hermitage (Figure 1, Sketch 2). This decision was supported by looming gender-related dangers in the forest that an aunty warned me about. During the monsoon weeks, countless public messages from the municipality on my phone urged me to stay inside and beware of mudslides (Figure 1, Sketch 5), implicitly forging a human allyship against the dangers of the volatile and hostile nonhuman environment.

With time, though, these 'hostile' *spaces* turned into *places* that I suddenly could relate to, appropriate, and therefore *subjectify*. This happened in conjunction with interviews and talks with residents, who helped me make sense of the spaces and showed me how to read and evaluate them, sometimes even in opposition to public warnings. While public messages on my phone drew the image of a dangerous landscape, residents contended that Noram had been spared from 'nature catastrophies' due to its fortunate geographic

'Even though aunty had warned me that there used to be a man that follows women up there, I tried to hike to the hermitage. As soon as I entered the trail though, the heat and humidity suddently densified. I felt like I was melting, sweating, spilling. When I saw a man resting on a stone, I decided to turn around…before it could maybe become weird..'

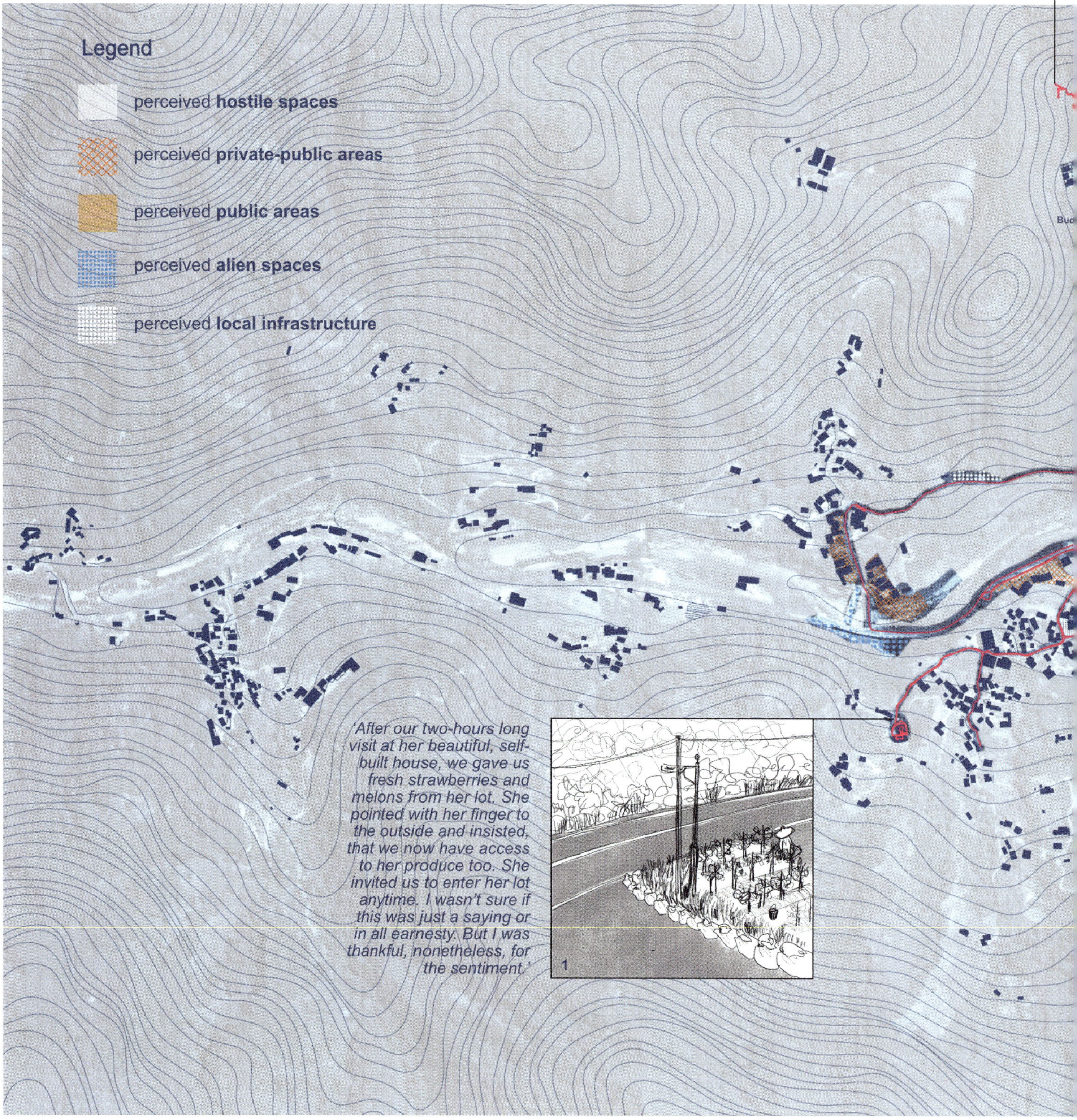

Figure 1 Temporal Point I MONSOON (Jae-Young Lee, 2024)

'The atmosphere was dense and humid. The silence was overbearing and the empty windows of the houses looked like eyes. And through them, I could see that some houses were completely ingested by plants.'

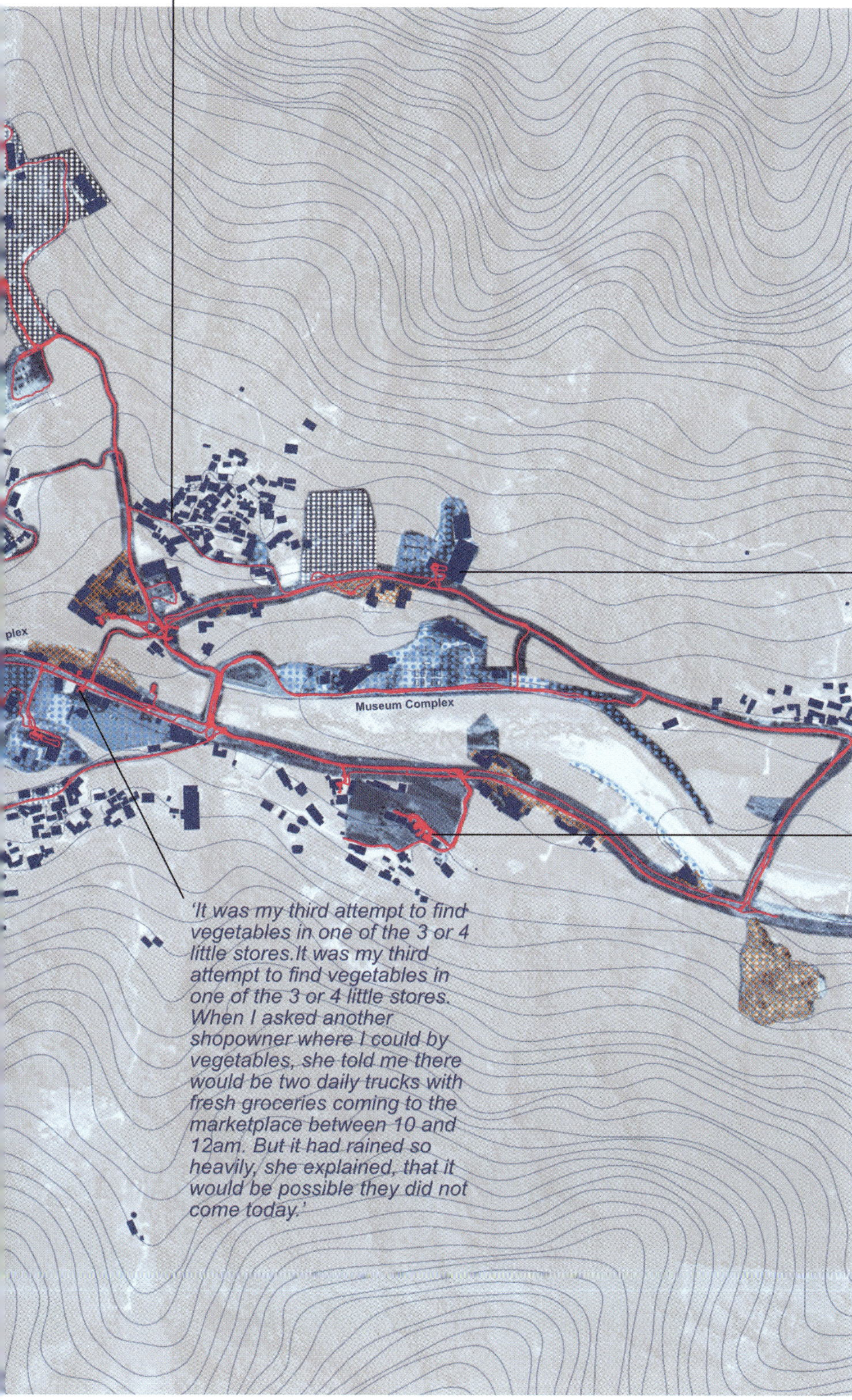

'It was my third attempt to find vegetables in one of the 3 or 4 little stores.It was my third attempt to find vegetables in one of the 3 or 4 little stores. When I asked another shopowner where I could by vegetables, she told me there would be two daily trucks with fresh groceries coming to the marketplace between 10 and 12am. But it had rained so heavily, she explained, that it would be possible they did not come today.'

'It was embarassing how comfortable I felt in the 7Eleven. I knew exactly how to behave there. There was even the only ATM and the counter lady immediately talked to me and told be that there are a lot of tourists from Russia at the moment.'

'I asked my host aunty, if it was safe to walk outside despite the messages. She chuckled a bit and said yes, and that this valley had always been safe. Because of its lucky geographical position.'

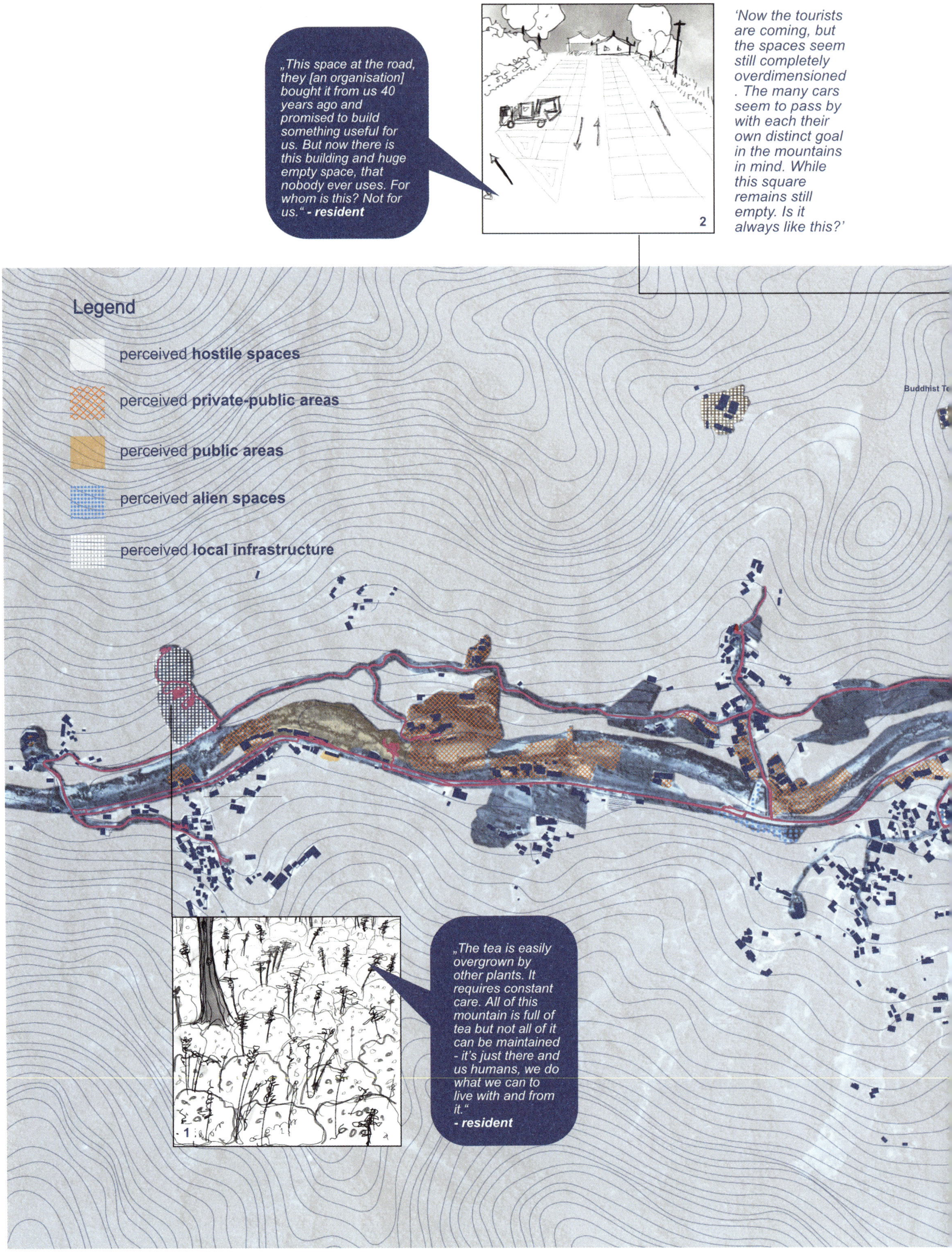

Figure 2 Temporal Point II NATIONAL HOLIDAY (Jae-Young Lee, 2024)

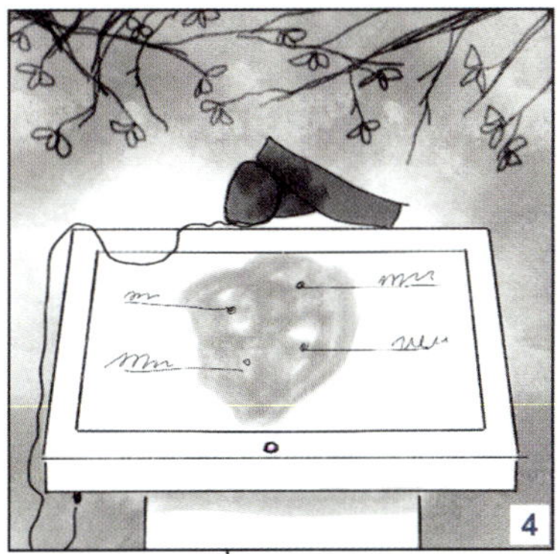

'The museum door was open, but noone to be found but a phone number. The exhibition was largely digital. I could even explore the whole region with Virtual Reality devices. Background music is linked to a motion detector, recording my every move. Everything seems to be digitalized and technologized through and through so that human contact can avoided.'

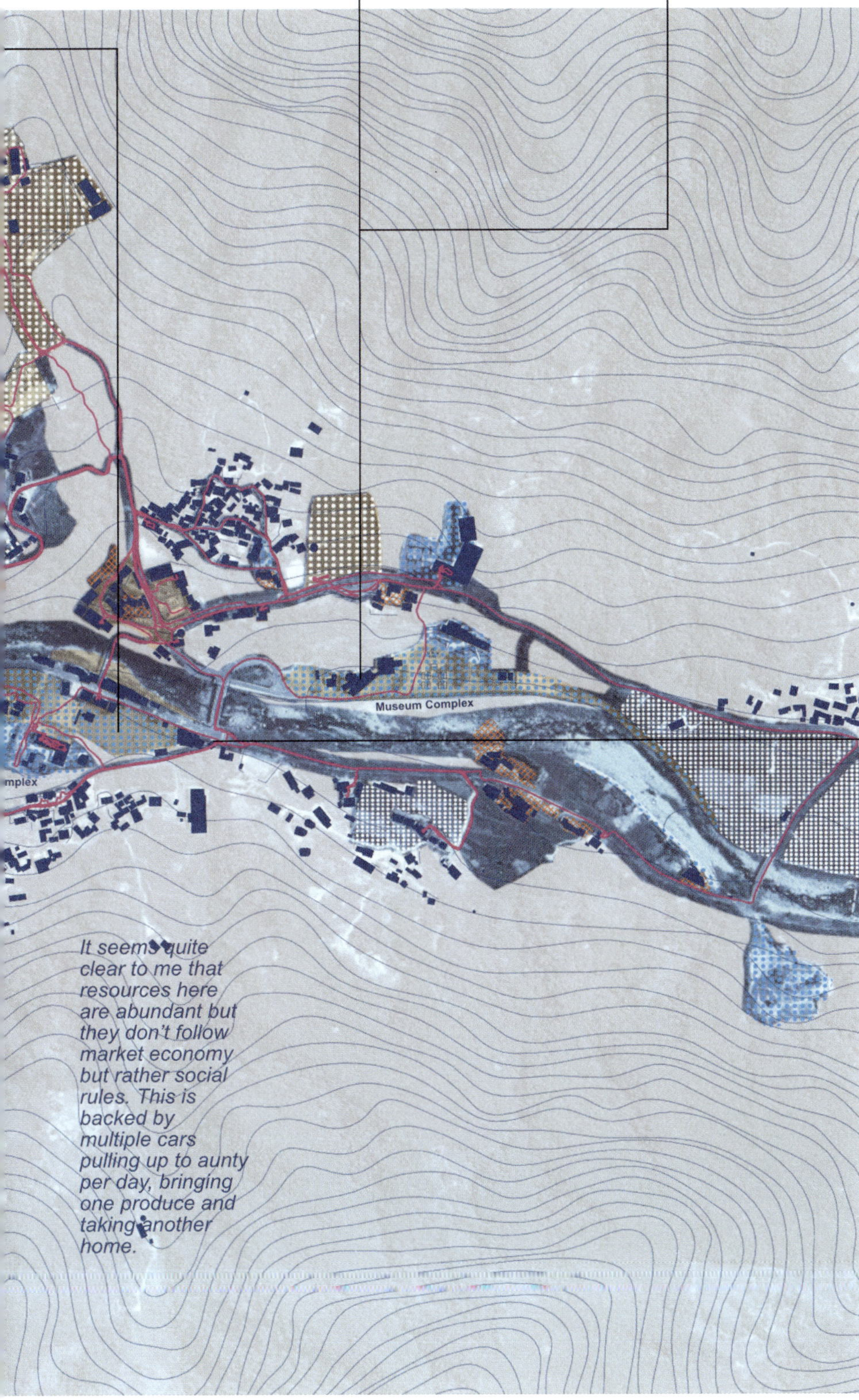

'The village entrance is steep and guarded by a traditional wooden gate. There's a banner saying that the village is against the construction. I wonder where they meet and I wonder who they are, 'the village'. But it feels great to see an immune-system of this place. At the same time, I don't feel comfortable to walk through the gate, I see no reason yet to intrude.'

It seems quite clear to me that resources here are abundant but they don't follow market economy but rather social rules. This is backed by multiple cars pulling up to aunty per day, bringing one produce and taking another home.

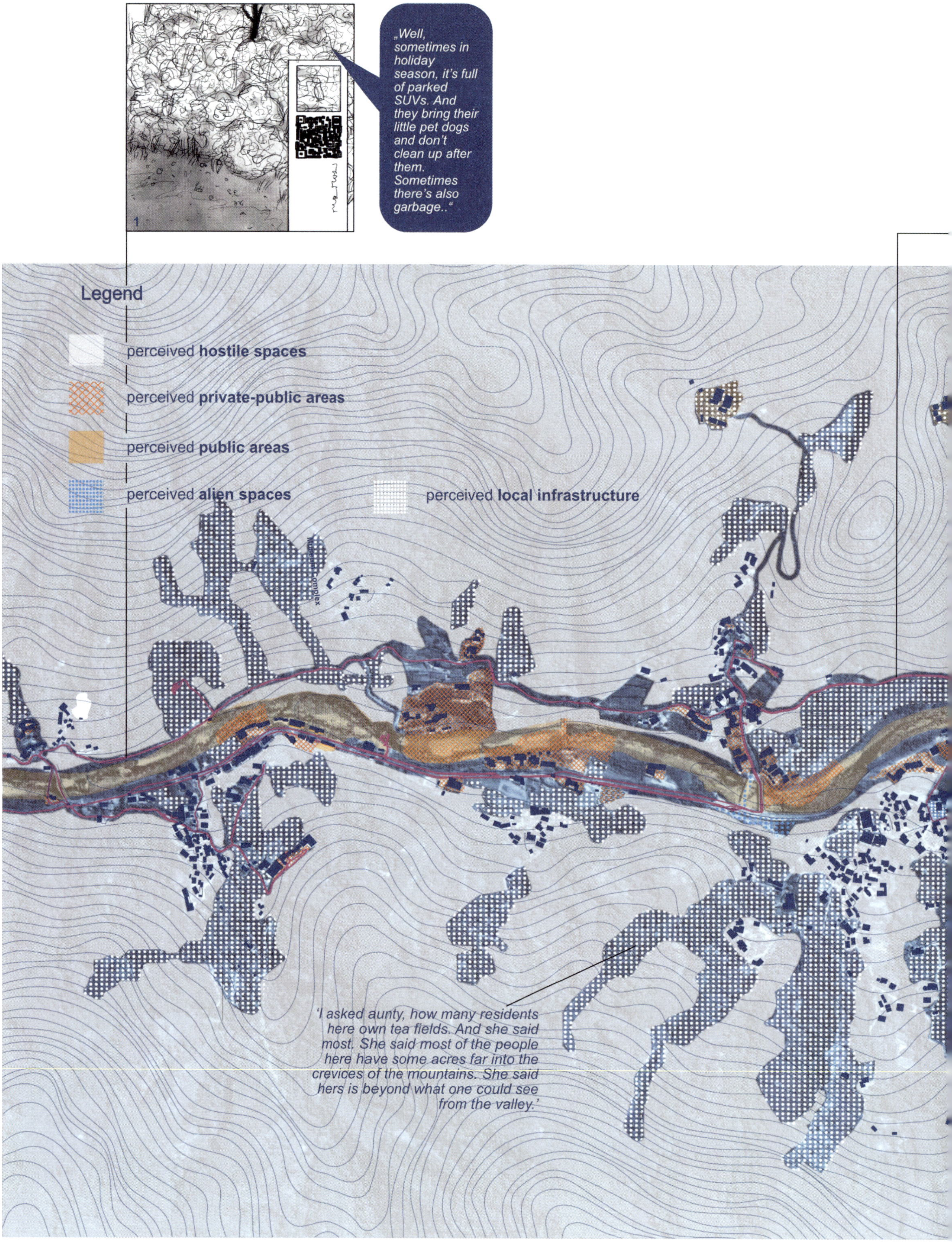

Figure 3 Temporal Point III LATE SUMMER (Jae-Young Lee, 2024)

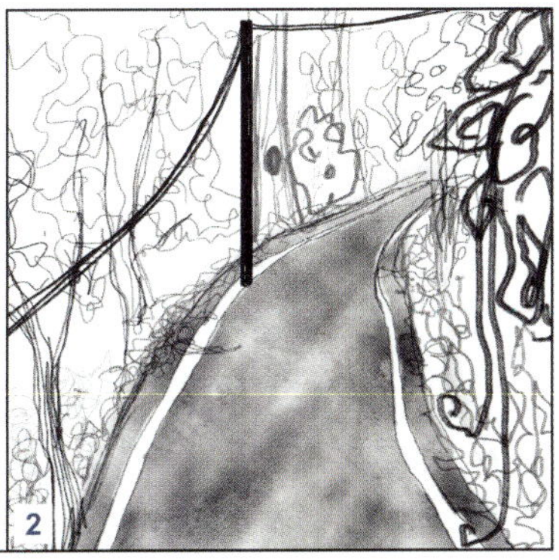

'Even though there is still dried blood and a bit of rotten smell along this road, I really prefer to walk this path. How has this feared, dark path with mysterious dead animals become my favorite?!
Now it feels so intimate and provides a cool breeze between the bamboo. I enjoy the peacefulness and the privacy from the all-seeing valley panorama. Finally, I am not watched.'

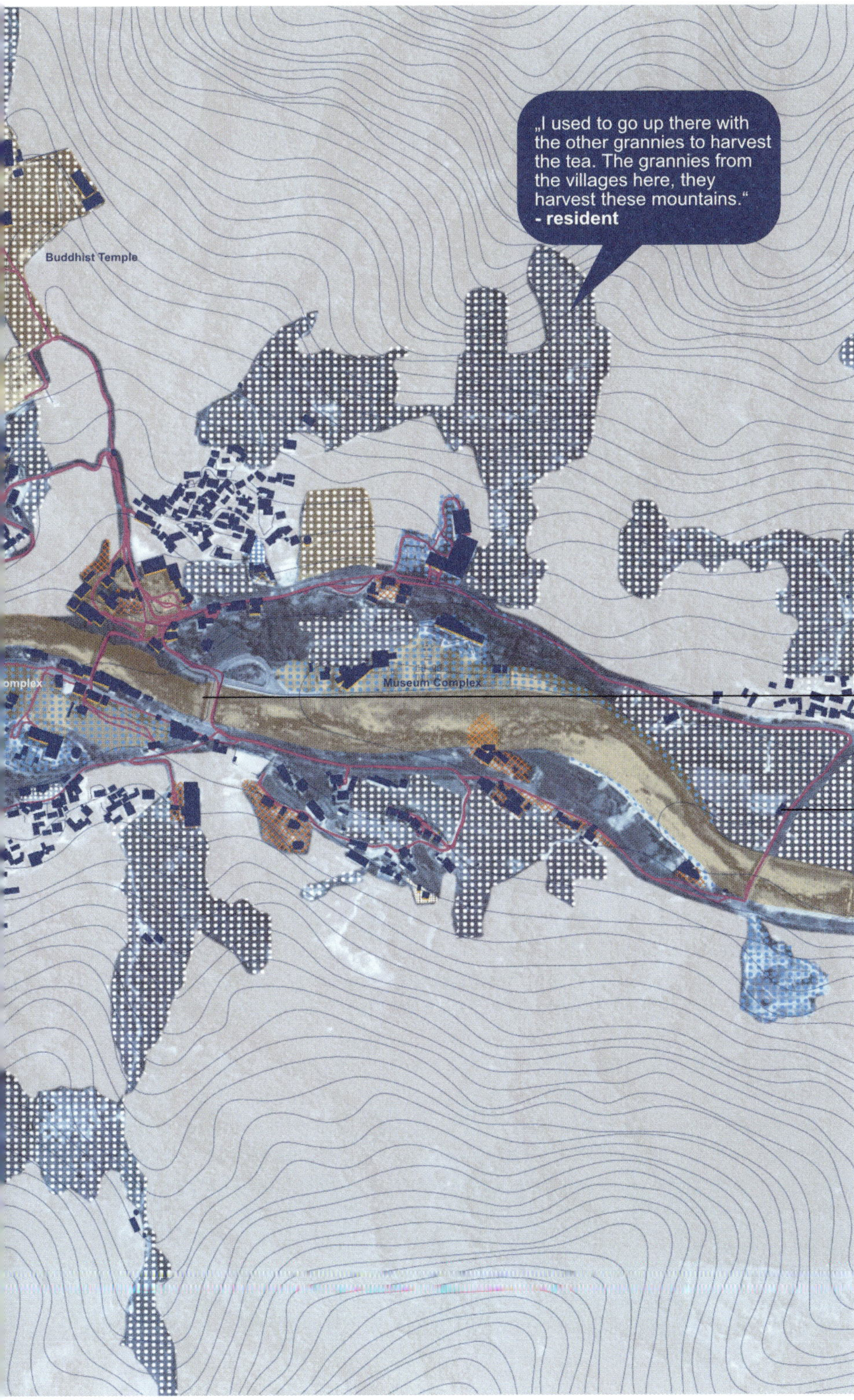

'Everywhere I look, there is people on the water. There seems to be truly no rule where you should or should not go. You just do it. When I asked Sajangnim, he politely responded: „Just find a nice spot and there will probably be a small path.'

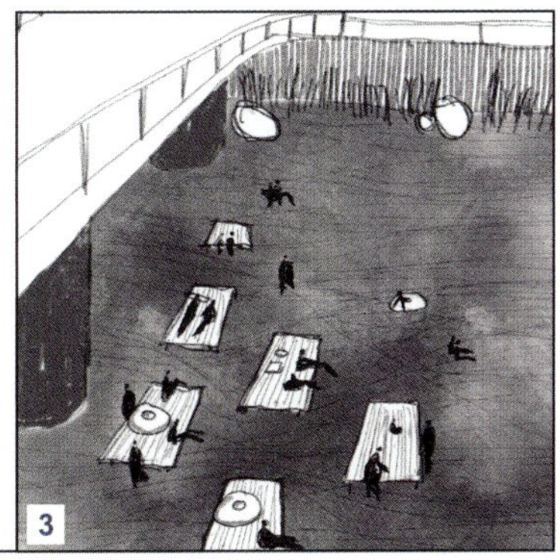

'As harvesting time approaches, I see more and more people in the fields. And it seems like auty's offer in the beginning was true. With host aunty, I witnessed how she yelled out: Big sister! Will you share some of your harvest with me? I have nice potatoes for you!'

configuration rooted in Korean spatial cosmology: mountain forest in the back and river to the front (배산임수). Hidden villages slowly became sites of chores, such as getting potatoes for an aunty. Empty houses became the former homes of mothers-in-law that the families did not have the heart to sell just yet. Additionally, with growing social networks and upon realising the tight social monitoring in place, formerly unsafe, shady paths hidden from the all-seeing valley panorama became places of refreshing privacy (Figure 3, Sketch 2). This new relational understanding also unfolded on a larger scale, detached from my bodily experience through residents' stories (Figure 2 and 3). The act of weeding a tea hill with a tea master illustrated the absence of clear demarcations between 'cultivated' and 'naturally occurring' flora, which I initially had perceived as an impenetrable border (Figure 2, Sketch 1). By listening to and exploring the practices and perceptions of residents in their familiar spaces, I gained insights that enabled me to understand them as lived landscapes of life and work (see Ingold 2000). Nonhuman actors and their spatial configurations that I had previously interpreted as spatial codes for hostile thresholds subsequently became mere material conditions that I had problematised due to my lack of spatial experience and knowledge and my own spatial 'programming'.

Public by Code vs. Public by Practice

'Public spaces', as a category, was used to sensitise places where I would feel legitimate and comfortable to dwell. They presented the direct opposite of the expansive 'hostile' places in the beginning. By employing the descriptive categories of 'alien' spaces and 'local infrastructure', two kinds of public spaces emerged: public spaces established by spatial codes that I could read and public spaces established by their appropriation and function as 'local infrastructures'.

Regarding typologies, large-scale asphalted squares with drawn-on demarcations and uniformly designed street signs presented in a spatial language were familiar to my current city life in Berlin (Figure 2, Sketch 2). Similarly, the 7-Eleven convenience store in the single high-rise hotel towering above the valley stood out as one of the most 'public' places (Figure 1, Sketch 4). As a global franchise, there were no codes that I had to figure out; the only required interaction was of a monetary nature. 'Public', in this sense, equated to anonymous interaction in a space facilitated by a financialised ruleset. Similarly, the vast number of uniform and designated public toilets, along with the considerable size of parking lots, seemed to constantly hold capacities for invisible masses. However, during monsoon season, this spatially conjured crowded image was rather juxtaposed—as museums had irregular opening times and personnel would often be absent, virtual reality devices assumed explanatory functions (Figure 2, Sketch 4). These large-scale structures, such as museums and squares, were moreover observed to congregate along the flattened landscape around the two smoothly asphalted highways along the river. Meanwhile, public spaces perceived as more local and even infrastructural to the residents could be explored at higher altitudes.

In recognising and learning these specific spaces, temporal dimensions played a significant role. Lonely benches high up in the hills and countless small vegetable lots that I perceived to be random and unstructured at first became infrastructures of local barter economies after a few weeks. Observations of subsistence farming, small-scall entrepreneurships, and bartering outlined these seemingly unstructured patches as an infrastructure that entangled humans with the land and each other. For example, aunties and uncles generously encouraged us to access their produce fields as soon as our relationship grew closer (Figure 1, Sketch 1; Figure 3, Sketch 4). Tea masters explained the necessity of aligning their lives with the seasonal rhythms of the tea plants, learned through biographical time that would even span generations (Figure 2, Sketch 1). At the lowest altitude, the river stood out as the most prominent public space that changed with the seasons. As the summer progressed, its rapid white water became crystal clear. At night, I curiously observed residents diving with headlights for fresh fish and snails to eat, sell, and/or barter. Portable outdoor furniture and sun canopies appeared in changing locations (Figure 3, Sketch 3). These places thus became public places through individual appropriation, depending on the spatial knowledge about the dynamics of shade, the pace of water, the stony undergrounds, and such.

In addition to this dynamic production of 'public' spaces, there were also cases in which appropriated public spaces would be superimposed by 'alien' spatial codes. As such, the narrow and hidden labour path of the tea hill had received a QR code from the municipality and had since become a tourist destination (Figure 3, Sketch 1). Residents recounted the presence of parked cars obstructing the pathways since then during summer season, as well as left-behind garbage and dog excrement. The path was now also adorned with little bridges in the recurring design language of the municipality. A wooden sign, glazed and smooth to the touch, peeked out between shrubbery, declaring, 'Place to take a picture' (author's translation).

Power Relations and Correlations: Spatial Learnings

Within this qualitative approach based on limited stories and fragments, the chosen categories of 'hostile', 'public', and 'private-public' spaces proved useful in exploring the spatial actors that constituted my discomfort and relief in this rural area. The superimposition of nonhuman dynamics on my bodily experience stood at the core of my spatial learning and subsequently encompassed and embodied the affective spatial layers of my own positionality. Encounters with the barter economy and residents' perceptions of 'critical' situations, such as typhoons, outlined a specific territorial and more-than-human metabolism (Perrotti 2022) in which agencies were directly negotiated by residents through their spatial knowledge and experience. This was contrasted with spaces that carried 'universal' spatial codes shaped by dislocated urban infrastructural knots, such as municipal or regional offices.

This distinction between residents' negotiations and universalised spatial mediation emerged at the heart of my own spatial learning and was mirrored in the difference between the spaces I initially perceived as 'public' or 'hostile'. The first map, with my initial perceptions of 'hostile' spaces and my comfort in 'alien spaces', illustrated my intrinsic search for universalised spatial codes that I could instantly read without spatial learning. Here, time emerged as a crucial vector in my autoethnographic spatial learning that informed negotiation processes between more-than-human agencies. Knowl-edge and experiences from human biographical temporalties were also entangled with the transgenerational duration of plant lives and seasonal changes. In contrast, 'alien' public spaces with their spatial codes seemed to command absolution in time and space. These spaces seemed to predefine their exclusive uses but also mediate relations between dust, dirt, and rain as well as humans, cars, and animals through their spatial form. They brandished clear demarcations and well-known typologies that implicitly curated and policed every action in space, ready for any contingency and exuding institutional and urban-centred authority. As illustrated in the maps, these spaces seemed to pierce through less-coded landscapes, like a pixelated pattern along the flattened areas around the river. This evoked processes of urbanisation as infrastructuralisation along valley plains (Valz Gris 2023), equally echoing the proliferation of urban experiences through 'standardised built environments and highly formatted management technologies and systems' (Simone 2023, 383). Consequently, the overwhelming presence of dynamic, time-sensitive, and less-coded spatial entities was perceived as unrationalised, uncontrolled, and therefore *unconsumable* at *first sight*. This inaccessibility seemed to constitute a threshold for immediate appropriation, an *'irrationality'* of presumably 'uncharted land' that was, as Markaki (2023, 222) writes, 'chaotic at first glance and chaotic and dispersed like its community'.

Against this backdrop, 'alien' spatial codes superimposing 'local infrastructures' seemed to align with accounts of the capitalisation of rural areas that replace experiential knowledge through epistemologies of rationalisation (Gerber 2014; Marglin 2008).

With urbanisation conceptualised as a 'proliferation of territories of experience' within capitalist spatial consumption (Simone 2023, 383), mapping my spatial learning processes sensitised concrete agents associated with imaginary urban control and its more-than-human counteragents. It was only through the residents' knowledge that such spaces became *subjectified places* that I could safely navigate with my body. In this respect, spatial learning seemed to constitute a threshold for the increasing encounters with spatial objectification that the residents associated with unregulated mass tourism and the resulting environmental degradation.

With the larger aim of researching the spatial perceptions and practices of a rural place, autoethnographic experiential mapping ultimately helped spatialise power relations and correlations between my own naturalised spatial codes and those I learned through residents. The categories and the choice of parameters, such as time and mode of transport, certainly require further scrutiny, as their extension would lead to new findings. However, exploring my own researching gaze highlighted the crucial role of more-than-human entanglements within contemporary classifications, such as rural and urban, layered with attributes, such as comfortable or even hostile. In this regard, questions of 'urban' and 'rural' become questions of control and power relations that shape our spatial experiences in more-than-human territories. This, along with the question of how the distribution of urban, universalised experiences might inform processes of extended urbanisation, remains to be examined further.

Acknowledgements

This article would not have been possible without the permission of the people of Noram, who shared their time, knowledge, and places with me. I owe it entirely to them that I can write this article with my name under it. I would also like to thank Sungwon Ryu for her patient explanations, translations, and analytical and kind mind.

Funding Statement

The research was funded by the German Research Foundation (Deutsche Forschungsgemeinschaft). Projektnummer 290045248 – SFB 1265.

Image Credits
All Figures by Jae-Young Lee

References

Arboleda, Martín. 2016. 'In the Nature of the Non-City: Expanded Infrastructural Networks and the Political Ecology of Planetary Urbanisation: In the Nature of the Non-City'. *Antipode* 48(2): pp. 233–251. https://doi.org/10.1111/anti.12175.

Brown, Katrina M., Frode Flemsæter, and Katrina Rønningen. 2019. 'More-Than-Human Geographies of Property: Moving Towards Spatial Justice with Response-Ability'. *Geoforum* 99: pp. 54–62. https://doi.org/10.1016/j.geoforum.2018.12.012.

Elwood, Sarah A. 2004. 'Experiential Learning, Spatial Practice, and Critical Urban Geographies'. *Journal of Geography* 103(2): pp. 55–63. https://doi.org/10.1080/00221340408978576.

Gerber, Julien-François. 2014. 'The Role of Rural Indebtedness in the Evolution of Capitalism'. *The Journal of Peasant Studies* 41(5): pp. 729–747. https://doi.org/10.1080/03066150.2014.921618

Gillon, Charles. 2014. 'Amenity Migrants, Animals and Ambivalent Natures: More-Than-Human Encounters at Home in the Rural Residential Estate'. *Journal of Rural Studies* 36: pp. 262–272. https://doi.org/10.1016/j.jrurstud.2014.09.005.

Hawthorne, Timothy L., Patricia Solís, Brittney Terry, Marie Price, and Christopher L. Atchison. 2015. 'Critical Reflection Mapping as a Hybrid Methodology for Examining Sociospatial Perceptions of New Research Sites'. *Annals of the Association of American Geographers* 105(1): pp. 22–47. https://doi.org/10.1080/00045608.2014.960041.

Ingold, Tim. 2000. *The Perception of the Environment: Essays on Livelihood, Dwelling and Skill.* London/New York: Routledge.

Korean Statistical Information Service. 2023. 'Population by Administrative District (Si/Gun/Gu) and Gender'. *Statistics.* https://kosis.kr/statHtml/statHtml.do?orgId=101&tblId=DT_1B040A3&vw_cd=MT_ZTITLE&list_id=101&seqNo=&lang_mode=ko&language=kor&obj_var_id=&itm_id=&conn_path=MT_ZTITLE.

Langran, Elizabeth, and Janine DeWitt. 2020. *Navigating Place-Based Learning: Mapping for a Better World.* Cham: Springer International Publishing. https://doi.org/10.1007/978-3-030-55673-0.

Lendvay, Márton. 2021. 'Watermelon Production as the Driver of Community Resilience: More-Than-Human Agency and the Transforming Rural Assemblage'. *Journal of Rural Studies* 86: pp. 376–385. https://doi.org/10.1016/j.jrurstud.2021.06.018.

Marglin, Stephen A. 2008. *The Dismal Science. How Thinking Like an Economist Undermines Community.* Cambridge, MA: Harvard University Press.

Marguin, Séverine, Dagmar Pelger, and Jörg Stollmann. 2021. 'Mappings as Joint Spatial Display'. In *Handbuch Qualitative und Visuelle Methoden der Raumforschung*, edited by Anna Juliane Heinrich, Séverine Marguin, Angela Million, and Jörg Stollmann: pp. 381–399. Bielefeld: transcript Verlag.

Markaki, Metaxia. 2023. 'Expropiation and Extended Citizenship. The Peripheralisation of Arcadia'. In *Extended Urbanisation. Tracing Planetary Struggles*, edited by Christian Schmid and Milica Topalovic, pp. 197–234. Basel: Birkhäuser.

McFarlane, Colin. 2011a. 'Assemblage and Critical Urbanism'. *City* 15(2): pp. 204–224. https://doi.org/10.1080/13604813.2011.568715.

McFarlane, Colin. 2011b. *Learning the City: Knowledge and Translocal Assemblage.* Chichester: Wiley-Blackwell. https://doi.org/10.1002/9781444343434.

Pavez, Isabel, Teresa Correa, and Javier Contreras. 2017. 'Meanings of (Dis) Connection: Exploring Non-Users in Isolated Rural Communities with Internet Access Infrastructure'. *Poetics* 63: pp. 11–21. https://doi.org/10.1016/j.poetic.2017.06.001.

Peake, Linda. 2016. 'The Twenty-First-Century Quest for Feminism and the Global Urban'. *International Journal of Urban and Regional Research* 40 (1): pp. 219–227. https://doi.org/10.1111/1468-2427.12276.

Perrotti, Daniela. 2022. 'Toward an Agentic Understanding of the Urban Metabolism: A Landscape Theory Perspective'. *Urban Geography* 43 (1): pp. 1–11. https://doi.org/10.1080/02723638.2020.1848760.

Powell, Kimberly. 2010. 'Making Sense of Place: Mapping as a Multisensory Research Method'. *Qualitative Inquiry* 16 (7): pp. 539–555. https://doi.org/10.1177/1077800410372600.

Powell, Kimberly. 2016. 'Multimodal Mapmaking: Working Toward an Entangled Methodology of Place: Multimodal Mapmaking as Emplaced Methodology'. *Anthropology & Education Quarterly* 47 (4): pp. 402–420. https://doi.org/10.1111/aeq.12168.

Pugh, Rhiannon, and Alexandre Dubois. 2021. 'Peripheries within Economic Geography: Four "Problems" and the Road Ahead of Us'. *Journal of Rural Studies* 87: pp. 267–275. https://doi.org/10.1016/j.jrurstud.2021.09.007.

Robinson, Jennifer. 2022. *Comparative Urbanism. Tactics for Global Urban Studies.* Hoboken: Wiley. https://onlinelibrary.wiley.com/doi/10.1002/9781119697589.fmatter.

Robinson, Jennifer, Fulong Wu, Phil Harrison, Zheng Wang, Alison Todes, Romain Dittgen, and Katia Attuyer. 2022. 'Beyond Variegation: The Territorialisation of States, Communities and Developers in Large-Scale Developments in Johannesburg, Shanghai and London'. *Urban Studies* 59 (8): pp. 1715–1740. https://doi.org/10.1177/00420980211064159.

Schmid, Christian, and Milica Topalovic. 2023. *Extended Urbanisation. Tracing Planetary Struggles.* Basel: Birkhäuser.

Simone, AbdouMaliq. 2023. 'When Extended Urbanisation Becomes Extensive Urbanisation'. In *Extended Urbanisation. Tracing Planetary Struggles*, edited by Christian Schmid and Milica Topalovic, pp. 381–399. Basel: Birkhäuser.

Sun, Shiwen. 2023. 'Transformation from Urban and Rural Planning to Territory Spatial Planning'. *Frontiers of Urban and Rural Planning* 1 (1): p. 2. https://doi.org/10.1007/s44243-022-00004-5.

Valz Gris, Alberto. 2023. 'Beyond the Boom. Genealogies of Corridor Urbanism in the Making of the Lithium Triangle, Argentina and Chile'. *Geoforum* 147: Article 103913. https://doi.org/10.1016/j.geoforum.2023.103913.

Woods, Michael. 2016. 'Territorialisation and the Assemblage of Rural Place: Examples from Canada and New Zealand'. In *Cultural Sustainability and Regional Development*, edited by Joost Dessein, Elena Battaglini, and Lummina Horlings, pp. 29–42. London: Routledge.

Flow Mapping

Lili Carr,
Alder Keleman Saxena,
Victoria Baskin Coffey,
Jennifer Deger,
Feifei Zhou,
Anna L. Tsing

Mapping the More-Than-Human in *Feral Atlas*

Introduction: *Feral Atlas* and Mapping the More-than-Human

What is 'more-than-human'? This term denotes a 'turn' in the social sciences, consolidated in the past decade or so, that sees 'the social' as a function of not just human thought and action but also the co-mingling of human lifeways with those of nonhumans. These fellow travellers can include not only other biological species but also non-living entities, or spiritual forces, all of which have the capacity to exert force and provoke reaction, bringing them into a 'social' frame[1]. For the purposes of this essay, a critical element of the more-than-human turn is that it places an emphasis not on drawing conceptual boundaries around certain configurations of the social, but on understanding how these configurations emerge out of relations. This includes relations among people, as well as those with and within the more-than-human world.

1 Some key readings in this area include writings by Haraway (2003, 2016) and Tsing (2013, 2015; Tsing et al. 2017). Readers may also wish to consider writings by Bennett (2010) and Mol and colleagues (Abrahamsson et al. 2015).

This chapter provides an example of how such relations might be mapped by drawing on our experiences of creating *Feral Atlas: The More-than-Human Anthropocene* (Tsing et al. 2021)[2]. This open-access digital project documents the role of nonhuman entities, which, sparked by interactions with human-built infrastructure, have developed new capacities that exceed the bounds of human foresight and control. *Feral Atlas* contains more than eighty 'field reports' on such entities, which include not only living species (such as Australia's invasive cane toad or carpets of kudzu in the southeast United States) but also non-living entities (such as anti-fouling paint on cruise ships, toxic mercury carried by wind currents, or 'feral' carbon dioxide). These 'feral' natures co-create the conditions we refer to, collectively, as the Anthropocene.

2 This essay draws heavily on conceptual work underpinning *Feral Atlas* and particularly on three essays originally published in that project by Carr (2021), Deger (2021), and Coffey (2021). Readers will note that, rather than restating the ideas in new words, we have in many cases drawn long quotes from these earlier essays. We encourage curious readers to refer to the original texts, both to explore the material in its context and for the deeper citational base that underpins them.

In *Feral Atlas,* the field reports were written by a diverse group of observers, including ecologists, historians, and anthropologists, alongside artists, filmmakers, Indigenous activists, and others. As a unifying design element, at the header of each report, in collaboration with the authors, we placed what we think of as a 'flow map' (hereafter referred to simply as a 'map'[3]), named as such to emphasise its focus on the dynamic *processes* that emerge out of many different kinds of relations. We conceptualised these maps as representing not space to be navigated but rather relational activity, as visual renderings of the spatiotemporal movements that characterise the feral dynamics described in the field reports. Put another way, our flow mapping experiments were animated by a spirit akin to this volume's call to 'hold together space and time in drawings' (Baxter et al., editorial, this volume).

3 *Feral Atlas*'s usage of the term 'flow map' differs from the term's usage elsewhere. For example, in geoinformatics, 'flow maps' are used specifically to depict the spatial movement of entities between fixed geolocations, which runs counter to our use of the term.

The kinds of movements that our flow maps show include blockages, agitations, transformations, combinations, processes, and entanglements—all types of relations that actively form Anthropocene worlds. When thinking about feral dynamics, we ask the following questions: What kinds of processes or relations matter in this particular instance? How might we make their force tangible to the viewer?

For the purposes of our project on the feral dynamics of the Anthropocene, what this exercise in mapping helps show is that more-than-human mapping requires different scopes, scales, temporalities, and foci than classic forms of humanist mapping. As Jennifer Deger writes:

> The challenge of the Anthropocene ... is to move beyond the universalizing implicit in a planetary approach, by attending closely to the intimate conjunctions of ecologies and social forces in specific places and processes, thereby becoming attuned to what can be learned there. To do this requires not only many different maps and multiple scales. It requires many different kinds of maps ... .
> (Deger 2021, emphasis added)

In attuning to these processes, we found inspiration in counter-cartography, particularly in Black and Indigenous counter-mapping[4]. Like these practices, mapping the more-than-human also requires stepping outside of a taken-for-granted view and showing things that are counterintuitive (or illegible) on the dominant 'grid'.

4 See, for example, Crampton and Krygier (2015), Harley (2013), Kinkaid (2023), and Roane and Hosbey (2019).

The following selection of maps aims to foreground the kinds of relationality that come into view when prioritising nonhuman agency in the transformation of land, sea, and air. The five included here accompany field reports in *Feral Atlas.* Within *Feral Atlas*, each is intended to help orient the reader within the feral 'flows and blockages' that arise in the place-specific entanglements of more-than-human processes with human-built infrastructure in the story being told.

Outside of *Feral Atlas*, these media do not necessarily function as maps. They do not obviously exhibit the features we might expect of more conventional mapping systems that represent points of interest in relation to geospatially defined terrain, such as a fixed scale, coordinates, orientation, frame, units of measure, and graphic projection. However, read as flow maps in the context of the *Feral Atlas*, each renders *new terrains* composed of relations stretching beyond the spatial and drawing lines of connection that situate us within the ecological processes and social forces in specific places and histories. Neither simply illustrative nor straightforwardly empirical, the choices of materials, genre, content, and style for each map speak with the accompanying field report and, more generally, with the *Atlas*'s argument as a whole.

Figure 1 American Bullfrog (Victoria Baskin Coffey, 2021)

As such, the spatial significances and temporal extents that each map gestures towards, as well as their framing, orientation, and perspective, are given depth by the accompanying field report. We encourage the reader to view these at https://feralatlas.org/.

Leakage and Spill: American Bullfrog

We begin with a drawing by Victoria Baskin Coffey (Figure 1), which introduces the field report on American bullfrogs (*Lithobates catesbeianus* or *Rana catesbeiana*) by wildlife biologists Gary Witmer and Nathan Snow (2020). This field report—and Coffey's drawing—exemplify the conditions of leakage and spill, two forms of uncontainability through which the Anthropocene takes shape. The failure to contain—or even attempt to contain—lively things is one way in which human-built infrastructures change their surrounding landscapes.

American bullfrogs refuse containment. Indigenous to the Northeast United States and the Great Plains, bullfrogs are invasive species in other parts of the world, including the Western United States and South America. In these regions, bullfrogs eat other frog species and displace local amphibian populations. They are also suspected of introducing diseases, such as the chytrid fungus *Batrachochytrium dendrobatidis (Bt)*, which is responsible for amphibian deaths of pandemic proportions worldwide (Kolby and Berger 2021; Snow and Witmer 2020).

Once introduced, bullfrogs are difficult to control. They are highly mobile, travelling up to a kilometre over land to find new territory, they have generalist eating habits (including eating the young of their own species), and they reproduce quickly and in great numbers, with some females producing as many as two clutches of 1000–4000 eggs per year. However, because the species that they displace are closely related and sensitive, methods for eradicating them are difficult to come by. Hand capturing is labour intensive and slow, and other methods, such as draining ponds, may negatively affect the local amphibians that bullfrog eradication would protect (Kolby and Berger 2021; Snow and Witmer 2020).

Coffey's map points precisely to these challenges. Here, bullfrogs sit within concrete pens modelled on commercial frog farms in Brazil. However, these lively creatures are having none of it, leaping from their inadequate pens in search of new terrain. Neither, for that matter, does the water the bullfrogs are penned in remain static, leaking—pathogens and all—through cracks in the enclosure to the water table below.

Regarding the process of mapping these actions, Coffey (2021) writes:

> Each scientific report, when read deeply enough, offers up a unique moment of poetic paradox. They describe designed human interventions upon the environment that animate nonhuman relationships in unexpected ways. These designed and undesigned consequences eventually stretch the world in different directions, ripping it into a brand new and devastating shape.

This is a map not of a world that is whole but rather of the processes that tear it apart.

Blockage: Colonial Railways in the Bengal Delta

Our second example, also drawn by Coffey, exemplifies the inverse of spill and leakage: ferality emerging from forms of blockage (Figure 2). The specific blockages here frame the field report by historian Iftekhar Iqbal (2021), which explores the uncontrolled proliferation of water hyacinth in the Bengal Delta.

As Iqbal describes, railways were introduced in Bengal in the eighteen-fifties to facilitate the export of rice and jute. However, while this tactic was initially successful at increasing revenue from exports, its long-term impacts were devastating to the agroecology of the delta region. Embankments and

Figure 2 Colonial Railways in the Bengal Delta (Victoria Baskin Coffey, 2021)

channels intended to make the land stable enough for railways had the effect of blocking the flow of rich silt from the Himalayas, allowing salt water to creep into the water table. Previously, this saline water would have been swept away by flows of fresh water. Although colonial officials were alerted regarding the extent to which these vastly changed agro-ecological conditions interfered with locals' ability to plant, they refused to remove the railways. By the mid-twentieth century, the region was dependent on imported rice—with the 1943–44 Bengal Famine among the dire consequences (Iqbal 2021).

Summarising this damage, Iqbal (2021) writes:

> ... the real question is why this rice surplus-producing and exporting region had become dependent on imports in only a few decades. Here, there are strong grounds to point to the role played by the railways, with their long arms of embankments, and the invasive plants that spread along them. What started as the most spectacular technological innovation of the time proved within a century to be a major vehicle of agro-ecological destruction.

Coffey's map of the Bengal Delta renders this violence palpable. In this representation, Coffey projects 'the forced stillness of water', juxtaposing water-coloured landscapes, made fertile by the gentle ebb and flow of tributaries and estuaries, with the harsh steel of railway tracks, rendered as solid, white, digital lines. It is through these layered lines that we witness the process of infrastructural ecocide as waters made stagnant fill with sediments and the water hyacinth itself, suffocating aquatic ecosystems and giving rise to human diseases. The map is intended to not only empirically demonstrate the impact of the railway network on the watery networks of life across a vast and vital region, but to evoke the *feel* of blockage, settling as a tightness in the viewer's chest (Coffey 2021).

Here, Coffey offers a consideration of what design may offer in the intimate relationship between map maker and reader. This insight suggests that we take our own bodies and feelings seriously as we cultivate forms of attention that ground us in situated encounters. Deger (2021) suggests thinking of maps as 'sensuous forms of coming to knowledge', teaching us to sense for ourselves in order to orientate, perhaps more intuitively, within the anthropogenic forces and more-than-human relationalities they engender or endanger. Doing so can perhaps provide us with another means by which to 'galvanize new forms of societal recognition and political response' (Deger 2021).

Emergence and Simultaneity: COVID-19

Made by Lili Carr over a period of twenty weeks during the spring of 2020, this map accompanies four field reports gathered for inclusion in *Feral Atlas* in the midst of the 'first wave' of the COVID-19 pandemic (Figure 3). The image provided here is a screenshot of the animated version on the *Feral Atlas* site, in which events emerge, one by one, slowly populating the page as the incidence and frequency of early-pandemic cluster events and its death toll rises.

Disease environments arise when human-made infrastructures promote crowding. Notice the repetition in programme and infrastructural typology that characterise, again and again, the cluster events reported: processing plant, factory, hospital, prison, ship. Sense perhaps a feeling of overwhelm as events pile up, rendering individual instances indiscernible, especially ones that, once reported, fade from sight.

The events on this map were compiled from multiple reports by local, national, and international news outlets. While the timing of the events in the animation corresponds to the pace of their appearance, there is no geospatial relation between each event popping into view and the plane of the map, only repetition and intensification in the volume and rate of occurrences. The infrastructures that take masses of people from one place to another and those that crowd people together in enclosed spaces, thus creating the

Figure 3 Coronavirus cluster events from 7 February to 27 June 2020 (Lili Carr, 2020)

31/05/2020 France
public housing
clusters accounting for 19% of all reported cases in France

15/05/2020 Münster, North Rhine-Westphalia, Germany
playground
20 cases in cluster

14/06/2020 Rome, Italy
squat
15 cases in cluster

)20 Christchurch, New Zealand
elderly care home
56 cases in cluster

29/05/2020 Singapore
international school
11 cases in cluster

22/05/2020 Skellefteå, Sweden
elementary school
18 cases in cluster

20/04/2020 Oise, France
high school
133 cases in cluster

27/05/2020 Münster, Germany
high school
7 cases in cluster

and, New Zealand
are home
in cluster

27/06/2020 Church End, Brent, UK
public housing estate
second largest cluster and highest age-standardised death rate in England and Wales

23/04/2020 Seoul, South Korea
call center
134 cases in cluster

Seattle, Washington, USA
fishing vessel
86 cases in cluster

Waynesboro, Georgia, USA
nuclear power plant
209 cases in cluster

2020 Waikato, New Zealand
lderly care home
15 cases in cluster

18/05/2020 Auckland, New Zealand
elderly care home
13 cases in cluster

18/05/2020 Auckland, New Zealand
elementary school
96 cases in cluster

26/05/2020 Fremantle port, Australia
cargo ship
6 cases in cluster

Maryland, USA
are home
in cluster

Baton Rouge, Louisiana, USA
bars
100 cases in cluster

Newton, Iowa, USA
wind turbine plant
131 cases in cluster

14/06/2020 Rome, Italy
hospital
109 cases in cluster

South Gate, California, USA
dairy processing plant
82 cases in cluster

20/05/2020 Pazardzhik, Bulgaria
car parts factory
22 cases in cluster

Island City, Oregon, USA
church
236 cases in cluster

n site
luster

08/05/2020 Matamata, New Zealand
bar
77 cases in cluster

Westland, Michigan, USA
psychiatric hospital
73 cases in cluster

Darien, Wisconsin, USA
foodstuffs processing plant
100 cases in cluster

rth Carolina, USA
n site
uster

27/06/2020 Seoul, South Korea
church
16 cases in cluster; second wave

Washington D.C., USA
psychiatric hospital
181 cases in cluster

Sherman, Texas, USA
foodstuffs processing plant
326 cases in cluster

Kansas City, Missouri, USA
paper manufacturing plant
201 cases in cluster

s County, Austin, USA
uction site
site clusters sites in the county

06/05/2020 Seoul, South Korea
nightclub
274 cases in cluster

Abilene, Texas, USA
industrial bakery
52 cases in cluster

26/05/2020 Stahnsdorf, Germany
mail distribution center
8 cases in cluster

15/04/2020 Pleven, Bulgaria
sewing factory
57 cases in cluster

Jerome, Idaho, USA
potato processing plant
67 cases in cluster

Singapore
struction site
cases in cluster

Singapore
construction site
83 cases in cluster

2020 Lima, Peru
sale market
ses in cluster

10/05/2020 Frankfurt, Germany
church
112 cases in cluster

Kankakee, Illinois, USA
fruit & vegetable processing plant
57 cases in cluster

Vancouver, Washington, USA
fruit processing plant
65 cases in cluster

17/06/2020 Beijing, China
wholesale market
137 cases in cluster; second wave

Ellensburg, Washington, USA
vegetable processing plant
64 cases in cluster

-Brabant, The Netherlands
sing plant
cluster

26/06/2020 Valencia, Spain
meat packing plant
6 cases in cluster

17/05/2020 Amsterdam, The Netherlands
choir practice
102 cases in cluster

Hazle Township, Pennsylvania, USA
Amazon fulfilment center
81 cases in cluster

17/05/2020 Central de Abasto, Mexico City, Mexico
wholesale market
ls; 90,000 workers; 300,000 buyers and delivery personnel visit each day

Storm Lake, Iowa, USA
pork processing plant
591 cases in cluster

nington, Minnesota, USA
k processing plant
749 cases in cluster

Loudon, Tennessee, USA
mushroom farm
57 cases in cluster

13/04/2020 Kuala Lumpur, Malaysia
mass gathering
1,705 cases in cluster; at least 10% of cases in Malaysia are linked to this cluster

19/06/2020 Lalla Mimouna, Morocco
fruit packaging plant
457 cases in cluster

)0 Coquitlam, British Columbia, Canada
getable processing plant
3 cases in cluster

25/05/2020 Queenstown, New Zealand
conference
39 cases in cluster

ogansport, Indiana, USA
neatpacking plant
900 cases in cluster

Guymon, Oklahoma, USA
pork processing plant
654 cases in cluster

25/06/2020 Huesca, Spain
fruit farm; worker dormitories
20+ cases in cluster

18/06/2020 Canary Islands, Spain
migrant paddle boat
14 cases in cluster

conditions for viral spread, are replicated indiscriminately across the planet. As a map of events without geospatial relations, this map hints at the space-time of rapid mass transit, through which distance is stretched, compressed, or erased altogether through networked connections, logistical coordination, and speed[5].

5 The Coronavirus flow map accompanies a series of short essays about the pandemic, which are linked together in *Feral Atlas* to evoke the sense of emergence and simultaneity that the curatorial team experienced in the summer of 2020. Among these, the stretching and compression of time and distance described here are most explicitly outlined in Liu's (2021) entry, which considers how 'just in time production' and the cross-continental linkages it entails laid the groundwork for the virus to spread to pandemic proportions.

This map brings into view the relation between human social dynamics, the present-day imperial and industrial spaces that spatially condition (and contain) these dynamics, and the conditions of crowding and conveyance that encouraged the novel coronavirus to emerge, not as a singular viral species or incidence of isolated infections, but as a pandemic. Yet, despite the spatial nature of these more-than-human entanglements, this map is structured temporally rather than spatially. The map uses clock time as its coordinates, where seconds structure the reporting of COVID-19 cluster events calibrated to the calendar date. Events are, of course, durational, and while the disease events themselves were empirically inaccessible to us, the relentless nature of the daily news cycle permitted our assumption that the most significant events (in terms of number of infections and number of deaths) would be reported. In our maps, disease events that pop into view remain for the duration of the infectious period of a typical COVID-19 infection (two weeks) before slowly fading from view. More recent events have already sprung up in their place.

This is a map made from the middle and from within, where there is no outside to step into from which to safely and detachedly survey the scene. It presents a snapshot of unfolding events—a partial view gleaned through the narrow lens of those news outlets whose reporting was freely accessible online from an IP address in northwest Europe in June 2020. It is a map of *reporting on* cluster events more than a map of cluster events *per se*, and it perhaps offers a useful reminder that mapping within the Anthropocene is always a mapping from within—often within conditions of profound disorientation—from an always partial perspective, with always partial knowledge. It is a process of finding one's bearings in scenes always in formation, taking them in as they unfold.

Nonhuman Map Makers: Dutch Elm Disease

This map (Figure 4) accompanies two field reports by two different authors, Clive Brasier (2021) and Susan Wright (2021), each describing the deathly spread of Dutch elm disease (*Ophiostoma sp.*) through elm tree populations across the United Kingdom and northwest Europe during the twentieth century. The two reports differ radically in genre, style, and perspective. Brasier's report follows the evolutionary development of different species of elm disease-causing fungal pathogens, while Wright's tells of her firsthand experience of the disease in its decimation of the elms that inhabited the land around her family's home in Essex.

This map offers a third perspective. It is a photograph of elm bark beetle breeding galleries in a piece of dying elm bark. Notice the web-like patterning of the tunnels. Various species of elm bark beetle are known to construct these galleries in springtime for laying their eggs, which hatch into larvae that feed in the galleries over the summer months. Once matured into young beetles, they disperse, feeding on healthy elm before finding a suitable tree in which to produce a brood. Dying trees infected with Dutch elm disease-causing fungal pathogens are particularly attractive for this. An elm with an accumulation of boring dust around its base is a sign that it is only a matter of a couple of seasons before young beetles will depart, carrying the deadly pathogens gleaned from the walls of their gallery nursery along with them. This is a found map, made by elm bark beetles, of the spatial distribution of elm disease pathogens dispersed through an elm tree's bark (Brasier 2001).

Rather than geolocate Dutch elm disease as it spreads and kills elms across a human-defined territory, this map draws our attention to the territory of an elm bark beetle in relation to an individual tree. It orients us to the spatial distribution of pathogens that are known to live in and around its galleries, while pulling us into the intricate nonhuman livelihood processes that entangle elm, beetle, and fungi. It does not take much of an imaginative leap to understand how elm beetle stowaways, with their accompanying fungal passengers, are transported overseas in the operations of the global timber trade.

This map is a found object—this piece of bark was prised off a dying tree, carefully turned over, observed closely, and documented. Our use of a photograph-as-map perhaps runs counter to our intention of mapping movement or flow, given that photographs capture a single moment, or 'slice' of time. However, as Sontag (2010) reminds us, it is precisely in the desire to freeze a single instance that we are aware of the unceasing dynamism and temporal nature of the world. Just as Sontag recognises the act of photography as a way to participate in a thing's inherent mutability and vulnerability, in *Feral Atlas*, we variously approached photographs not only as documentary or a form of witnessing from afar but also as a way of emotionally and intuitively engaging with the multiform registers of the worlds we are entangled in (Coffey 2021; Carr 2021).

This map is also a material artefact from the field. As something observed and gathered, it is thick with the situated context of place. However, as an orienting device, it is perhaps most usefully thought of as a form of mapping practice, where close observation and attendance, not in the abstract but *in situ*, offer situated, visceral encounters. It raises questions of method and what it takes to orient oneself, with one's feet on the ground and engaging one's bodily senses in the apprehension of processes and materialities in place. After all, this kind of map could never be charted from a distance, through a telephoto lens, or through an earth observation system, however closely one zooms in.

Intimate and Sensuous: Underwater Noise

'How might one use sensory methods to encourage—to mediate—a collective reorientation, without necessarily having to be there?' (Deger 2021). Maps that engage our sensory capacities more fully offer one such way.

This map is a sound recording that accompanies a field report by S. B. Blackwell (2021), who describes how industrially produced underwater noise rips into oceanic socio-spatial relations in ways that we are only beginning to apprehend. Blackwell guides us into ongoing research around the impact of the increase in volume and the extent of human-produced underwater noise and its effect on marine life, particularly on marine mammals, such as bowhead whales, whose lifespan can exceed 100 years. A companion report by Matthew Buttacavoli and Geoff McPherson (2021) explores the impact of rising levels of anthropogenic noise on marine life in the Great Barrier Reef. They discuss the relative lack of critical scientific and public attention that underwater noise receives compared to other pollutants, such as marine plastic, and suggest that a possible reason for this may be that humans are often underwater in places where noise is not, away from shipping channels. They also explain how human anatomy limits our ability to sense changes in the marine soundscape, a contextual 'deafness' that privileges and upholds a distinctly human perspective.

Now listen to the audio map.[6] Attune yourself to the gentle whine of an engine overlaid through the wash of underwater movement; hear the temporal measure of the echosounder as it charts the ocean floor, pulse by pulse. Become aware of the sonic texture that fills the space and beyond our human audible range, registering intimate nonhuman goings-on under water. These sounds tell of complex socioecological relations, even though we cannot parse them into individual sources and constituent parts.

However, our ability to attune, notice, and observe when we are no longer able to see is less limited than you might think. *Feel* the violent rupture across

6 Audio map available at https://feralatlas.supdigital.org/poster/there-is-no-escaping-underwater-noise

Figure 4 Elm bark beetle breeding galleries in a piece of dying elm bark. The pathogens Ophiostoma ulmi and O. novo-ulmi live in and around the galleries (Clive M. Brasier, 2001)

this sonic fabric as the airgun releases its charge—this technology is designed to prospect for hydrocarbon sinks deep beneath the ocean floor. *Feel* the force of the shift in ambient pitch, quality, and volume as a boat's engine cuts through and silences an otherwise expansive polyphonic environment.

As we stretch our hearing and attune our attention to registers beyond the visual, this map helps us orientate within a realm where we are not. It stretches the limits of our hearing, just as it stretches the limits of our understanding. Just as certain frequencies will always remain out of human audible range, so will our abilities and capacities to know. This does not mean that we cannot notice the more-than-human processes and relations that shape these worlds. Blackwell draws a parallel with this audio and an instinctive revulsion felt when confronted with a leaf blower while conversing with a colleague on a street corner. By expressing the force of Anthropocene processes across diverse registers, such a map viewed from afar can perhaps offer some version of being there, rendering the more-than-human relations and forces in those places and their histories more fully sensible and, indeed, palpable (Deger 2021).

Conclusion

An animating argument of *Feral Atlas* is that there can be no single perspective on the Anthropocene, for there is no collective 'we', human or more-than-human, who experiences the Anthropocene in the same ways and at the same costs. The challenge, then, of the Anthropocene, as many insist[7], is to reject tendencies towards universalisation—God's eye views[8] and planetary totalisations—by attending closely to the human/more-than-human relations and forces that shape processes and ecologies in specific places, even as such relations may encompass planetary scales.

7 See, for example, Demos (2017).

8 See Haraway's (1988) discussion of 'the God Trick'.

The collection of 'flow' maps in *Feral Atlas* is an attempt at a method of gathering many different kinds of maps, expressing many different kinds of relational dynamics, and holding them together and in conversation in the framework of a digital atlas. The navigational architecture of *Feral Atlas* is designed in such a way as to invite readings across and between them, linking them to each other and hoping that new perspectives and new paths may emerge in their intersections, even as they refuse to add together neatly. As Coffey (2021) writes:

> The stories you will encounter here have happened and are happening. They are in motion and at variable scales. They ask you to zoom in, and to zoom out, and to zoom in again, and while you do that you must allow your peripheral vision to grasp the relationships that make this potential possible. That is where the Anthropocene is located. The Anthropocene is in that motion: it is alive at the edges of things.

For *Feral Atlas*, map making and map reading are situated activities of being-with as much as thinking-with. They require navigating lines of attention and inquiry and following signposts beyond the bounds of the map into the unknown. Deger (2021) writes:

> It is in this implicit tension, between the known and the unknown, the charted and the off-the-charts, that maps claim their special, more-than-empirical appeal. They are orientating devices. They compel us forward. They enable us to seek out new paths. They hold us, frame perspectives, make the path onwards visible. Or at least imaginable.

As we follow along, noticing and drawing disparate entities into relation and laying lines of connection, our maps generate other maps, just as they move us and shape us, helping us see things differently.

This relational approach to more-than-human mapping productively converges with tendencies in the design world, particularly with recent conversations emphasising materiality, in combination with renewed calls for a more critical and politically engaged approach to visual representation amid anthropogenic environmental crises. Designers are learning to be attentive to the materials they use and to follow them as they flow through spatial, temporal, and sociopolitical processes that have bearing on their practices. However, this shift in focus presents challenges to conventional mapping and representational methods due to how nimbly material processes shift across scopes and scales and because tracking these processes requires attention to the work of agents—both human and nonhuman—whose perspectives are not typically centred in design-oriented representational practices. One way of approaching these challenges, as this volume advocates, is to integrate spacetimes by insisting on approaches that work across multiple scales and registers and that eschew linear notions of historical progress (Baxter et al., editorial, this volume).

In *Feral Atlas*, we included not only archival and cartographic maps but also maps as audio recordings, maps as films, maps as paintings, maps as songs, and more, as a way of opening up representational possibilities and exploring their potential. Perhaps different kinds of maps, or learning to read maps differently, can help us open new worlds when we attune to our senses beyond sight alone. Perhaps such maps can help us become oriented within complex ecological relations that escape easy articulation and, in doing so, provoke new forms of recognition and response.

In these maps and methods, *Feral Atlas* offers only a starting point—hopefully one that opens new lines of inquiry into mapping more-than-human relations, map by map by map.

Image Credits

Figure 1: Victoria Baskin Coffey
Figure 2: Victoria Baskin Coffey, with map adapted from Iftekhar Iqbal (2010)
Figure 3: Lili Carr, Screenshot
Figure 4: Clive M. Brasier (2001)

References

Abrahamsson, Sebastian, Filippo Bertoni, Annemarie Mol, and Rebeca Ibáñez Martín. 2015. 'Living with Omega-3: New Materialism and Enduring Concerns'. *Environment and Planning D: Society and Space* 33 (1): pp. 4–19. doi: 10.1068/d14086p.

Bennett, Jane. 2010. *Vibrant Matter: A Political Ecology of Things*. Durham: Duke University Press.

Blackwell, Susanna B. 2021. 'There Is No Escaping Underwater Noise'. In *Feral Atlas: The More-than-Human Anthropocene*, edited by Anna L. Tsing, Jennifer Deger, Alder K. Saxena, and Feifei Zhou. Redwood City: Stanford University Press.

Brasier, Clive M. 2001. 'Rapid Evolution of Introduced Plant Pathogens via Interspecific Hybridization'. *BioScience* 51 (2): p. 123. doi: 10.1641/0006-3568(2001)051[0123:REOIPP]2.0.CO;2.

Brasier, Clive M. 2021. 'Introduced Pathogens Can Evolve Rapidly, Increasing Their Virulence'. In *Feral Atlas: The More-than-Human Anthropocene*, edited by Anna L. Tsing, Jennifer Deger, Alder K. Saxena, and Feifei Zhou. Redwood City: Stanford University Press.

Buttacavoli, Matthew, and Geoff McPherson. 2021. 'Cargo Ship Noise Jeopardizes Ocean Life'. In *Feral Atlas: The More-than-Human Anthropocene*, edited by Anna L. Tsing, Jennifer Deger, Alder K. Saxena, and Feifei Zhou. Redwood City: Stanford University Press.

Carr, Lili. 2021. 'Flow Maps and Moving Drawings'. In *Feral Atlas: The More-than-Human Anthropocene*, edited by Anna L. Tsing, Jennifer Deger, Alder K. Saxena, and Feifei Zhou. Redwood City: Stanford University Press.

Coffey, Victoria Baskin. 2021. 'Mapping the Anthropocene: A Letter to the Human Navigator'. In *Feral Atlas: The More-than-Human Anthropocene*, edited by Anna L. Tsing, Jennifer Deger, Alder K. Saxena, and Feifei Zhou. Redwood City: Stanford University Press.

Crampton, Jeremy W., and John Krygier. 2015. 'An Introduction to Critical Cartography'. *ACME: An International Journal for Critical Geographies* 4 (1): pp. 11–33. doi: 10.14288/ACME.V4I1.723.

Deger, Jennifer. 2021. 'You Are Here'. In *Feral Atlas: The More-than-Human Anthropocene*, edited by Anna L. Tsing, Jennifer Deger, Alder K. Saxena, and Feifei Zhou. Redwood City: Stanford University Press.

Demos, T. J. 2017. *Against the Anthropocene: Visual Culture and Environment Today*. Berlin: Sternberg Press.

Haraway, Donna. 1988. 'Situated Knowledges: The Science Question in Feminism and the Privilege of Partial Perspective'. *Feminist Studies* 14 (3): pp. 575–599. doi: 10.2307/3178066.

Haraway, Donna. 2003. *The Companion Species Manifesto: Dogs, People, and Significant Otherness*. Cambridge: Prickly Paradigm Press.

Haraway, Donna. 2016. *Staying with the Trouble: Making Kin in the Chthulucene*. Durham: Duke University Press.

Harley, John Brian. 2013. 'Maps, Knowledge, and Power'. In *The Iconography of Landscape: Essays on the Symbolic Representation, Design and Use of Past Environments, Cambridge Studies in Historical Geography*, edited by Denis Cosgrove and Stephen Daniels, pp. 272–312. Cambridge: Cambridge University Press.

Iqbal, Iftekhar. 2010. *The Bengal Delta: Ecology, State and Social Change, 1840–1943*. Basingstoke: Palgrave Macmillan.

Iqbal, Iftekhar. 2021. 'In the Bengal Delta, the Anthropocene Began with the Arrival of the Railways'. In *Feral Atlas: The More-than-Human Anthropocene*, edited by Anna L. Tsing, Jennifer Deger, Alder K. Saxena, and Feifei Zhou. Redwood City: Stanford University Press.

Kinkaid, Eden. 2023. 'Counter/Cartographies: An Editorial Introduction'. *You Are Here: The Journal of Creative Geography* XXIV: pp. vi–ix.

Kolby, Jonathan, and Lee Berger. 2021. 'Global Trade in Frogs Has Led to Catastrophic Amphibian Decline'. In *Feral Atlas: The More-than-Human Anthropocene*, edited by Anna L. Tsing, Jennifer Deger, Alder K. Saxena, and Feifei Zhou. Redwood City: Stanford University Press.

Liu, Andrew B. 2021. 'Is COVID-19 a "Just-in-Time" Pandemic?'. In *Feral Atlas: The More-than-Human Anthropocene*, edited by Anna L. Tsing, Jennifer Deger, Alder K. Saxena, and Feifei Zhou. Redwood City: Stanford University Press.

Roane, J. T., and Justin Hosbey. 2019. 'Mapping Black Ecologies'. *Current Research in Digital History* 2. doi: 10.31835/crdh.2019.05.

Snow, Nathan, and Gary Witmer. 2020. 'Introduced American Bullfrogs Are Invasive'. In *Feral Atlas: The More-than-Human Anthropocene*, edited by Anna L. Tsing, Jennifer Deger, Alder K. Saxena, and Feifei Zhou. Bloomington: Stanford University Press.

Sontag, Susan. 2010. *On Photography*. New York: Picador.

Tsing, Anna L. 2013. 'More-than-Human Sociality: A Call for Critical Description'. In *Anthropology and Nature*, edited by Kirsten Hastrup, pp. 27–42. London: Routledge.

Tsing, Anna L. 2015. *The Mushroom at the End of the World: On the Possibility of Life in Capitalist Ruins*. Princeton: Princeton University Press.

Tsing, Anna L., Nils Bubandt, Elaine Gan, and Heather Anne Swanson, eds. 2017. *Arts of Living on a Damaged Planet: Ghosts and Monsters of the Anthropocene*. Minneapolis: University of Minnesota Press.

Tsing, Anna L., Jennifer Deger, Alder K. Saxena, and Feifei Zhou, eds. 2021. *Feral Atlas: The More-than-Human Anthropocene*. Redwood City: Stanford University Press.

Wright, Susan. 2021. 'I've Lost the Smell of Elm Dust'. In *Feral Atlas: The More-than-Human Anthropocene*, edited by Anna L. Tsing, Jennifer Deger, Alder K. Saxena, and Feifei Zhou. Redwood City: Stanford University Press.

Anne-Marie Willis, Tony Fry

Future-Risk Mapping

Between Time and Space

Introduction

1 This chapter revisits Willis (2012), bringing it to risk mapping, as elaborated by Hay et al. (2024).

This chapter focuses on 'mapping as designing'[1]—not the use of mapping within design, but the stronger claim of mapping (map making and map using) as a designing force. Over the last three decades, the widespread use of the global positioning system (GPS) and associated technologies that facilitate real-time navigation has transformed mapping as an idea and practice. It has become dynamic, emphasising movement rather than fixed locations. This is not just a matter of technological progress; rather, it needs to be considered ontologically as formative for new ways of being in time and space that inevitably displace other modes of doing and knowing. The traditional, static spatial map will also be shown to have designing agency, although of a different order in terms of the centrality of cartography for colonialism's projects of extraction and occupation.

Surveys and site maps conceal time, yet spaces are mapped as precursors to change. Thus, maps are catalysts for ending the time of one thing and inaugurating the time of what is to come. This is disastrously evident now, with sea levels rising, coastlines changing, and polar ice caps disappearing. New deserts will form; forests will continue to be lost; rivers will dry; settlements will, and are, being abandoned.

Although computer cartography (digital mapping) enables continuous revision (thus providing an ability to map in time), with an ease in changing the scale and greater flexibility in the use of map imagery, it still remains within the same cartographic paradigm. In contrast, and in contemporary enviro-climatic circumstances, risk mapping now overshadows all mapping. However, risk mapping needs to be developed into a more ambitious practice, one that maps dynamic ongoing 'change events' and their interrelations rather than just climate and geography. We have named this more ambitious practice 'mapping in time', which, we argue, needs to be brought together with 'designing back from the future'. This is concretised by a discussion of some speculative and exploratory projects undertaken in Australia.

We should add that we come to mapping via design and are thus less concerned with positioning our work within mapping as a specialist area of knowledge. Rather, our approach—epistemologically, functionally, and politically—is to see that the plurality of the practice of mapping now needs to be strategically taken up and used to indicate viable futures that respond to the forces of defuturing that we humans and other-than-humans in our differences are travelling towards (Fry 2020).

The Mapping Explosion

In the twenty-first century, mapping has moved well beyond spatiality and physical geography. It has diversified technologically and migrated across many disciplines: cultural mapping, mapping social networks, mapping the inside of the body, mapping the genome, mind mapping, cultural mapping, conversation mapping ... the list goes on. Some of this is weak—mapping as metaphor, or re-describing modes of visual representation, such as diagramming, as 'mapping'. The distinction between map, diagram, plan, and visualisation starts to blur. Mapping, beyond the representation of geophysical and geological space and conditions, thereby becomes a label for analysing and giving visual form to data.

More significantly, this mapping explosion has been driven by the availability of vast amounts of all kinds of data that lend themselves to being analysed and presented visually. Additionally, there has been the proliferation of mobile devices that are precisely locatable via satellite, thus allowing for mapping in real time and collapsing together what were once distinct, temporally separate activities of mapping and navigating.

Ways of Being in Space

Mapping, as an idea and practice, has become dynamic, emphasising movement rather than fixed locations, processes rather than places, and time as well as space. The static map prefigured actions in/on worlds (discussed in more detail later), but it was not instantaneous. It did not operate in real time, as satellite/GPS navigation does. What this now normalised technology did was shift mapping away from representation and interpretation towards embodied experience. Map and territory became one—a realisation of the Borgesian image of the map that exactly corresponds with the territory (1:1), except that the correspondence is *temporal not spatial*—the blue or red dot sliding across a screen in synch with the tracked target, such as your car. While the map is still there as part of the navigation system, as an electronic image of lines, place names, and landmarks, it is mostly superfluous to navigation, as it has been displaced by audio instructions (e.g., 'at the next intersection turn left ...' etc.).

While the technology and applications of real-time mapping have advanced and can be investigated in terms of functionality, it is worth considering it ontologically, as it has created new ways of being in time and space. In this example, the satellite-tracked driver or pedestrian no longer needs to 'figure out' where they are and where they are going because their device does that for them. What does it mean to be lost now? Is a sense of direction redundant? As geolocation devices and applications have rendered anywhere instantly navigable and useable, does this expand opportunities for encountering (and learning from) the unfamiliar, or does it reinforce the familiar because it can be so easily located in the unfamiliar? As these technologies continually enable 'more', the impression is one of opening up—more information, more options, more accessibility, more transparency, more visibility. This obscures that what is going on is an induction into a particular way of knowing and being in the world—one that has become 'indispensable', displacing other ways of knowing and being in the world. Here is the ontological designing of technology as it designs its users (Willis 2006).

These observations of real-time mapping are not an expression of nostalgia for static, hard-copy maps. The traditional static map also has designing agency. It induced particular modes of knowing and acting, as a subject capable of mentally correlating lines and markings on a two-dimensional diagram to the landscape of streets and buildings through which body moves. Historically and culturally, static maps also design subjectivities as they prefigure certain kinds of movements—sightseeing, scenic drives, history walks, and so on (wayfinding as a design subgenre). The point here is that, in the shift from static maps to mobile, instantaneous mapping, there is a continuity of designing agency. The difference is in what is being designed.

The Map Is Not Neutral

Historically, map making was indispensable to colonialism's project of conquest and control, and the map was central to the very concept of the nation-state and its operation[2]. Map making is always directive, interested, never neutral; it makes sense of and creates orders as a prelude to taking control. Maps generate movement.

John Pickles (2003) and other critical geographers, such as Denis Cosgrove (1999), Trevor Barnes, and J.B. Harley (1992), have emphasised this constitutive agency of maps and mapping. The epistemological and strategic power of cartography has also been recognised in critical studies of colonialism, such as by Thongchai Winichakul (1988, as cited by Anderson 1991: 175), who argued against the common sense idea of a map merely representing something that already existed: 'A map anticipated spatial reality, not vice versa ... a map was a model for, rather than a model of, what it purported to represent .... It had become a real instrument to concretise projections on the earth's surface'.

Essentially, conventional mapping reifies space and power. This is most evident in the history of Western colonialism, wherein European nations claimed power over space as their forces of colonialisation created a map of territorial space. Their mapping began at the edge (coast or border;

2 The straight lines that define the borders of many African nations have been the enduring legacy of the European imperial nations' carve-up of Africa at the Berlin Conference of 1884–1885. These 'lines of agreement' were drawn from a European political perspective, paying no heed to the fracturing of tribes, communities, topographies, etc., and they have been the source of a good deal of the conflict in Africa that still continues.

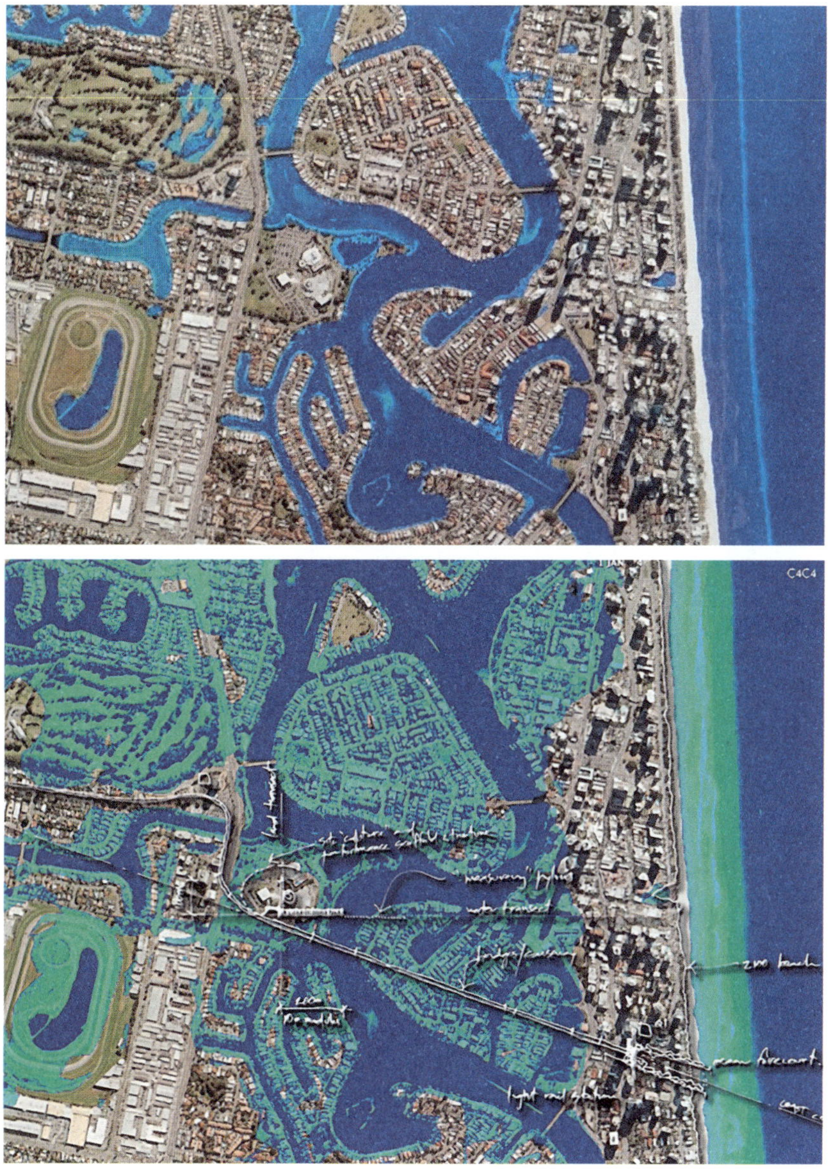

Figure 2 Gold Coast, Queensland 2010 and projection for 2100: hypothetical but plausible mappings of future losses due to sea-level rises (Tony Fry and Jim Gall, 2010)

geophysical barrier). Mapping the interior was indivisible from claiming and extending their power. In contrast, Indigenous people mapped towards the edge from their situated experience of reading the land and its physical and posited symbolic features. The space and time of their lifeworlds were taken away by the power of the map and what underscored it, as denoted by the geography of their subjectification and subsequent ontological transformation of being colonised spatially and existentially.

Mapping, Prefiguring, Planning, and Designing

Maps prefigure movement. Defining and locating features of territory, opening it up for certain kinds of occupation or use—this is their designing power. Conventionally, surveying and map making are considered as prior to, and separate from, the prefigurative acts of planning and designing. However, according to Denis Cosgrove, historically, the plan preceded the map. This was the case for so many cities, ancient and modern, and especially fortress and colonial cities. Additionally, geometry—the radial axis and the grid—historically underpinned both cartography and modern urban form, that is, the plan of the city (Cosgrove 2006).

Mapping and planning most often occur simultaneously. A site is surveyed, and the survey produces the information from which to make the map. The map opens up the possibility of a plan. Or, maybe the mapping has only occurred because the intention is to make a plan.

What is a plan? A plan always looks to the future. It enfolds action and change. A plan is a map projected forward in time. It says, 'this is what will be', or 'this is what is here now, and this is how it will be at some point in the future.' A plan is a design. This closeness of mapping, surveying, planning, and designing shows that maps are always interested. This is not to deny that a map really shows topographical features; by walking over the site, it can be verified that there is a line of trees here, a road there, and so on. But we need to ask: why is this site being mapped? What is the impositional intent? Mostly, it is an intention to occupy or make use of—to extract minerals; clear a habitat for farming; subdivide as real estate; build a road, a railway, a port.

The Site Map

A site map is a mapping of the already obliterated. Within the design professions, the move from mapping to designing is very fast. Here, mapping is a preparation for designing, and it can be done cursorily (e.g., by identifying 'site constraints' on what is to be built), or it can be open to the unexpected (thereby modifying what is to be built). More sensitive, fine-grained mapping can inform the direction and character of the finalised design, but it is unlikely to result in leaving things as they are, because the mapping itself is already framed by a prior decision that change will occur.

From Space to Time

Historically, a spatial paradigm ruled surveying and map making (and the activities that follow, such as the building of roads and railways). To the surveyor and cartographer in the service of capital or a colonial power, space was limitless—there to be mapped, claimed, surveyed, occupied, and built upon. The traditional survey map conceals time. It plots out locations in space, as if time is of no consequence, as if what it describes is permanent. Yet, spaces are mapped as a precursor to change; thus, maps were (and are) catalysts for ending the time of one thing and inaugurating the time of what is to come. Here, we speak of time as the endurance of something, rather than clock time.

Mapping Uncertain Futures

Now, 'in our time', time has come to the fore in mapping. This is evident in the visualisations of the most crucial issue of our time: climate change, especially the global warming scenarios, issued regularly by the Intergovernmental

Panel on Climate Change (IPCC)—those maps that show global temperatures rising in tandem with increasing carbon emissions. Associated with this, over the last thirty years, thousands of scientists have been mapping that which is disappearing. They are mapping the shrinkage of glaciers and polar ice caps; the changing distribution of species in habitats; the rates of clearing of vegetation; the dieback of trees; variations in stream flow, soil temperature, rates of decline of fish species, and trends in coral bleaching. Their mapping cannot keep pace with the speeding of loss and change.

There is a tragic connection here. This urgent *temporal mapping* is occurring now because of centuries of *spatial mapping* that conceived of the Earth (land and sea) as a boundless space to be occupied and a bounty to be reaped. In other words, mapping as a neutral spatial practice (or misrecognised as such) has been an agent of the taking away of time. *This, in the largest sense, is what the map has designed.*

The implication is that risk mapping now overshadows all mapping. What, then, are the implications for a critical practice of designing that is aware of this?

The answer is that critical designing will be 'designing in time'. What this means is projecting forward in time—picturing how things are changing as the very forces of change are unfolding—and then designing back from this projected future. This is, in fact, a practice of prefiguring prefiguration—not with absolute certainty, not with a finished 'design product' in mind, but with flexibility and responsibility.

Designing in time is predicated on the recognition that, while designers, architects, and engineers have had an extraordinary ability to bring wonderous things into being—be they structures, artefacts, technologies, or systems—they have been bad at identifying the consequences of what they created. This failure is writ large in all that is unsustainable, from the creation of ongoing climate change to the loss of biodiversity and the vast ocean islands of plastic waste. Designing in time, a method that counters this tragic creative flaw, was first used in 2007 as a submission to an international concept design competition for a sustainable city for 50,000 people[3]. Our approach was based on Boonah, an existing town in South East Queensland, to establish a notional life-span of the 'to be designed' and then undertake a 'narrative mapping' of all the potential 'impact events' (environmental, social, economic) over the assigned design life. The entry, titled 'Boonah Two', involved research on likely future scenarios in order to speculate on responses and generate design propositions—all presented as plausible narratives and visualisations unfolding over time in a specific place[4].

As a method, 'designing back from the future' demands research to combine the discoverable and the speculative. So, while it does not claim certainty, it does extend the ability to take responsibility. It has been used in further projects, including several on remaking and relocating cities and higher education projects including a response to sea-level rises in the Pearl River Delta in Hong Kong (see Fry 2015, 2017; Willis 2014). Most recently, it influenced *Relocating Australian Communities at Risk: Strategies and Action in Time* (Hay et al. 2024). There is no exportable template for the method. Rather, the principle needs to be remodelled for each different project, although the three narrative lines of the initial 2007 'Boonah Two' concept are usually appropriate to employ (Figure 1). They are 1) a year-by-year listing of impacts (when projected to arrive), 2) an indication of the responses of those who are impacted by them, and 3) design responses to both (as designed in the initial act of creation, or programmed into its life cycle).

Recognising mapping as an act of construction, not representation, it now needs to be strategically taken up and used to indicate viable futures that respond to the forces of defuturing that we humans and other-than-humans in our differences are travelling towards. This position informed 'Gold Coast 2', undertaken in 2010, which, like the Boonah project, was a deliberately noncompliant entry for a design competition. The brief called for ideas for a master

3 'Building a Sustainable World: Life in the Balance', a competition organised by the Royal Institute of British Architects/USA. The entry, 'Boonah Two', was the work of Tony Fry and architect, Jim Gall. Its content and method is elaborated in a case study by Fry (2009).

4 While 'Boonah Two' was noncompliant with the competition rules at the presentations at the Los Angeles Architectural Museum, where thirteen of the fourteen shortlisted entrants from nine different countries presented master plans, our narrative map was awarded second place.

plan for a civic centre site on Queensland's Gold Coast. Instead, the entry (by Tony Fry with Master of Design Futures students and architect Jim Gall; Figure 2) began by considering the future of the Gold Coast based on the IPCC's long-term projections of sea-level rises, with these informing proactive responses presented as visual and narrative mappings over future decades, including the staged relocation of much of Gold Coast City, and adaptive new uses for what might remain[5]. The entry was awarded 'Supplementary Winner', and the designers were invited to address the planners of the Gold Coast City Council.

5 A good account of 'Gold Coast 2' can be found at the website of Nora Kinnunen, one of the designer participants: https://norakinnunen.com.au/?page_id=95

Mapping Future Complexity

As a result of a changing climate, there is a discernible drift from the relative stability of the global geophysical to a condition of growing instability. Sea levels are rising; coastlines are changing; polar ice is disappearing. New deserts will form; forests will continue to be lost; rivers will dry; settlements, towns, and cities will and are being abandoned or destroyed; populations will or already are being displaced; borders will blur and be transgressed; conflict will increase.

This is not a dystopic projection but a recognition that climate change is irreversible. It is based on the IPCC evaluation of thirteen climate models marking increased warming until 2300, ending with a still upward trend, including an increase of 2.7°C by 2100. This is in the context of an interglacial cycle of warming that will last several thousands of years.

Currently, risk maps are created by governmental and private agencies to identify places that are or will be exposed to risks such as flooding, sea-level rises, increasing temperatures, and fire. Mostly, they are concerned with a small number of factors and are limited in their capability to model complex emergent risk. However, risks are not discrete; they overlap and impact each other and are variably dynamic. Current data, images, and their presentations do not deliver sufficient evidence of intersecting and compounding levels of risk across climatic, environmental, cultural, social, and economic situations. The complexity of what urgently needs to be mapped now can be briefly indicated by registering two perspectives: the outer and inner worlds of our being. The outer perspective is what can be seen, experienced, and mapped as a dynamic mapping in time. This can be elaborated on by the following categories and examples:

The global: Consider, here, the equator, cutting through the Indonesian archipelago, Central Africa, and the north of South America. Its heat will draw a death line. The Nigerian city of Lagos, just north of the equator, is projected to have a population of 80 million by 2100. Currently, it is a place to flee to. It will become a place to flee from.

The regional: This can be exemplified by the Indonesian capital of Jakarta, a sinking delta city with a population of more than 11 million. It is exposed to threats from flooding and sea-level rises, as well as pollution, earthquakes, and overpopulation. It is a delta city, built on a low, flat alluvial plain prone to flooding during monsoon season. Over 40 per cent of Jakarta is below sea-level. The city has sunk 4 metres in the past twenty years—a situation made worse by groundwater extraction. As a result, the government decided to move the capital to East Kalimantan, Borneo. Indonesia is a nation of islands that sprawls over more than 5,000 kilometres, and Kalimantan was chosen, as it was more central than Jakarta. Only the formal community deemed functional will move; those who remain will be left to their fate. The new capital, named Nusantara, will cause major environmental problems in Kalimantan. Its construction is underway, and this has meant clearing vast areas of flora- and fauna-rich tropical rainforest. Additionally, the new city and its population will add to the problems for the Dayaks, the Indigenous population, who, over many years, have been forcibly removed from their traditional areas to make way for palm oil plantations, paper production and mining sites. The entire project is due to be completed by 2045.

The local: This can be illustrated by Marble Bar, a small mining town in Western Australia. Its population is being burnt out of existence. Over a period of 160 days in 2022, the lowest temperature was 37.5°C, the highest over 50°C. Another example is Lismore, a small city in New South Wales, 734 kilometers northeast of Sydney with a population of 45,000. In 2021, it flooded three times as a result of extreme weather events; this happened again in 2022. In the flood of March 2022, the water rose by 14.4 metres. This was the worst flood in Australian history and effectively destroyed the city. Buildings were simply washed away.

The inner perspective recognises that climate is a massive force of ontological change. Tens of thousands of years ago, it was a key factor in distributing our species around the planet. It was an evolutionary force. As such, it determined the colour of our skin, our diet, our forms of shelter, what we did or did not wear, and our psychology. We—that is, globally, all of us—are now at the beginning of the second moment of planetary global redistribution.

To conclude, internal and global mass relocation will change where and how we live. Many agricultural systems will fail, nations will fall, major cities will be lost, borders will be redrawn, and conflict is certain to increase. We will change and likely start to fragment as a species. Here are three crude scenarios: that of the abandoned, the privileged techno-immune, and, in between them, the pluriversality and bricolage of the adapted masses. What is certain is that ontologically, we will not stay as we are.

Signs of these trends are already present. How they will unfold and with what relational consequence remains unclear. What is evident is that we are all becoming unsettled. This will increase not only physically but also psychologically. Nation-states will be under enormous stress, as will natural resources, and enviroclimatic and biophysical impacts will increase. Plus, the risk of conflict will be high.

The degree of severity of this situation will be determined by prefigurative and preventive action taken over the next few decades, if not centuries. As the IPCC has made clear, climate change is irreversible. In this context, a new type of proactive mapping needs to be developed. It will involve dynamic risk mapping that outlines situations where risk is a moving process and where the impacts of climate change interact with and exacerbate drivers of insecurity and instability. Such dynamic mapping would allow risks to be recognised and preemptively engaged before they materialise. It is an approach that has strategic value for communities, industry, and policymakers, offering more comprehensive projections of emerging, interrelated risks and able to inform strategic actions such as relocation planning.

Acting in time (the medium and with a state of urgency) is absolutely critical. Such action cannot simply be instrumental. It demands action upon our 'selves' to change our modes of being in the world—achieved by transforming the conditions that constitute our ontologies and habitus. This implies a situated transformation of our ecology of mind (our mode of knowing), ecology of image (our related mode of seeing), and the practices of worldly construction. What we, in our differences, do, what we make, how we live—all will ontogenetically determine what we become. This recursive process has brought us to this moment, and it will determine our future.

Image Credits

Figure 1: Tony Fry and Jim Gall, produced for Boonah Two competition entry, 2007

Figure 2: Tony Fry and Jim Gall, both visualisations produced for 'Gold Coast 2' competition entry, 2010

References

Anderson, Benedict. 1991. *Imagined Communities*. London: Verso.

Cosgrove, Denis, ed. 1999. *Mappings*. London: Reaktion Books.

Cosgrove, Denis. 2006. 'Carto-city'. In *Else/Where: Mapping: New Cartographies of Networks and Territories*, edited by Janet Abrams and Peter Hall. Minneapolis: University of Minnesota Design Institute.

Fry, Tony. 2009. *Design Futuring: Sustainability, Ethics and New Practice*. Sydney: UNSW Press.

Fry, Tony. 2015. *City Futures in the Age of a Changing Climate*. New York: Routledge.

Fry, Tony. 2017. *Remaking Cities: An Introduction to Urban Metrofitting*. London: Bloomsbury.

Fry, Tony. 2020. *Defuturing: A New Design Philosophy*. London: Bloomsbury.

Harley, J.B. 1992. 'Deconstructing the Map'. In *Writing Worlds: Discourse, Text and Metaphor in the Representation of Landscape,* edited by Trevor Barnes and James Duncan. New York: Routledge.

Hay, Naomi, Tony Fry, Roslyn Prinsley, John Handmer, Jeremy Mather, Milica Muminovic, Noam Maitless, Mathew Gill, Owen Crofts, Jessica Van Son, and Liam Taylor. 2024. *Relocating Australian Communities at Risk: Strategies and Actions in Time*. Canberra: ANU Institute for Climate, Energy and Disaster Solutions, Australian National University. https://iceds.anu.edu.au/files/Relocation%20Issues%20Paper.pdf.

Pickles, John. 2003. *A History of Spaces: Cartographic Reason, Mapping and the Geo-Coded World.* New York: Routledge.

Willis, Anne-Marie. 2006. 'Ontological Designing'. *Design Philosophy Papers* 4 (2): pp. 69–92.

Willis, Anne-Marie. 2012. 'The Ontological Designing of Mapping'. *Design Philosophy Papers* 10 (2): pp. 95–104.

Winichakul, Thongchai. 1988. *Siam Mapped: A History of the Geo-Body of Siam*. PhD Thesis, University of Sydney.

Willis, Anne-Marie. 2014. 'Designing Back from the Future'. *Design Philosophy Papers* 12 (2): pp. 151–160.

Angela Million,
Simten Önen,
Katrin Schamun

Gamified Mapping

Investigating Children's Well-Being in Space and Time

Introduction: Exploring Well-Being from Children's Perspectives

Children's well-being is deeply intertwined with the spaces they inhabit, navigate, and co-create, making the exploration of these socio-spatial dynamics vital for understanding their lived experiences (Castillo et al. 2024). This paper investigates how gamified collective mapping can serve as an innovative methodological tool for uncovering and representing these dynamics. Specifically, we ask the following: how can gamified collective mapping enable children to express their perceptions of well-being while capturing the socio-spatial and temporal complexities of their environments?

Collective mapping is widely recognised for its significant benefits in negotiating and producing collective understandings and outcomes (Cortés-Morales et al. 2022). Its action-enabling approach aligns with the need to take children and youth involved in research seriously (Million 2021). However, conducting research with children and youth involves navigating stringent ethical and legal frameworks, particularly concerning privacy, anonymisation, and data protection. These challenges highlight the importance of adapting research methodologies to meet ethical standards while amplifying the voices of young participants.

The methodological foundation of this paper is rooted in Massey's (2005) conceptualisation of spacetime as dynamic and relational. Massey (2005, 59) argues against 'conceiving of space as a static slice through time', emphasising instead that space is 'a simultaneity of stories-so-far' (Massey 2005, 9). This underscores her view of space as dynamic and interconnected. This lens informs our approach, emphasising the multiplicity and co-creation inherent in collective mapping. By employing a gamified approach, we aim to reconfigure the spatiality of research participation, in which children's narratives intersect and shape the collective mapping space.

This chapter outlines the potentials and challenges of collective mapping, particularly in linking children's well-being to socio-spatial contexts and space-time dynamics. Drawing from the interdisciplinary research project *Children's Well-Being in Socio-Spatial Contexts* (WIKK*I) (Million et al. 2024), we explore how children co-construct their well-being in extracurricular learning environments. The project critiques the adult-centred view of well-being (Ben-Arieh 2006; Fattore et al. 2017; International Child Development Centre 2016; World Vision 2013, 2018) and instead foregrounds children's perspectives and their capacities to co-construct spaces of meaning.

Through a gamified mapping approach (Brkovic Dodig and Groat 2021) inspired by our research setting—a children's centre in Berlin—we developed a participatory method to capture children's perceptions (Million et al. 2024). This chapter details the methodological choices underpinning this, including the adaption of mapping methods to the specific setting, the integration of visual methods for data analysis, and the anonymisation of sensitive data. We conclude by reflecting on the benefits and challenges of using a gamified collective mapping approach and its implications for future research and methodological advancements.

Research Setting, Ethical Standards, and Legal Ambiguities

The children's centre used as the setting for this research was an integral and dynamic environment that provided a fertile ground for engaging with children and studying their well-being. Situated in a vibrant urban district of Berlin, the centre is accessible by public transport and is characterised by its open green spaces. The centre's architecture combines functionality with safety, offering both enclosed and open spaces that support group activities and individual exploration. Conversations with the educational staff and the children revealed a diverse mix of socio-economic and cultural backgrounds, including the children of first- and second-generation immigrants. Most of the children were bilingual or trilingual, speaking various languages from the Middle East, Western Europe, and Africa, while a minority spoke only German.

Children's centres in Germany are influenced and guided by a social pedagogy committed to viewing children from the perspective of the subject, meaning that children are supported in their 'here and now' (Chassé et al. 2003). The goal is to enhance societal access and participation opportunities by providing spaces for specific groups and needs in terms of special assistance and times (Reutlinger et al. 2010). The pedagogical approach in these centres typically follows two strategies: structured variety and free self-organisation (Kessl and Reutlinger 2022). The children's centre in our research employs both, offering structured activities and allowing space and time for individual initiative. This creates a flexible and responsive environment that balances structured support with freedom for personal development.

The centre's role as a protective and nurturing space requires stringent ethical considerations, particularly regarding consent, privacy, and the impact of research activities on the children's routine. Ensuring that all research activities were transparent, voluntary, and conducted with minimal disruption to the children's daily activities was crucial. This meant that children were not forced to participate in the research and could opt out at any time during the gamified collective mapping sessions.

Additionally, anonymising sensitive data is essential to maintaining confidentiality and trust. While ethical standards are rigorously enforced and integrated into research frameworks internationally through established procedural ethics, obtaining approval from an ethical review board in Germany is not mandatory. German legal regulations do not significantly emphasise ethical issues but focus intensively on protecting individual privacy through strict anonymisation of personal data. In this context, research practices often require written consent, typically validated by a guardian's signature, underscoring the formalities involved (Christmann et al. 2018). While European legislation assumes that minors are able to give consent, this legal framework presents challenges for conducting research in walk-in facilities frequented by children who may arrive unaccompanied or come from diverse linguistic backgrounds. Our project adopted an approach that allowed children to express their consent independently and enabled them to withdraw from participating in the research and our gamified mapping activities at any time—a choice that a few children did indeed make. This flexibility addresses the ongoing legal ambiguities regarding a child's capacity to give informed consent as much as it accepts the fact that children do have a voice of their own (for more information, see Fischer and März 2024).

While anonymising personal details helps protect participant confidentiality, our research team also addressed the challenge of ensuring that the visual analysis and dissemination of children's well-being in maps containing geographic locations and markers were carefully considered. These considerations highlighted the need for a nuanced approach that balances ethical accountability with the integrity of spatial research. The setting, along with the ethical and legal aspects, significantly influenced our processes for data collection, analysis, and dissemination.

Method of Gamified Collective Mapping

Our key approach was to engage the children themselves as experts on their own needs. For our qualitative study, a combination of various ethnographic methods (Clark 2017) was carried out to capture diverse narratives, experiences, and understandings of what well-being means to children. Data collection and analysis were conducted in two phases between 2022 and 2023, encompassing five steps (Figure 1).

Step 1: Production of Game Set

Our game was developed based on data gathered in the first phase of data collection at the children's centre. Ethnographic observation methods (DeWalt and DeWalt 1998) were employed, as well as interviews and go-alongs with the children (Kusenbach 2008) conducted in the nearby urban environment.

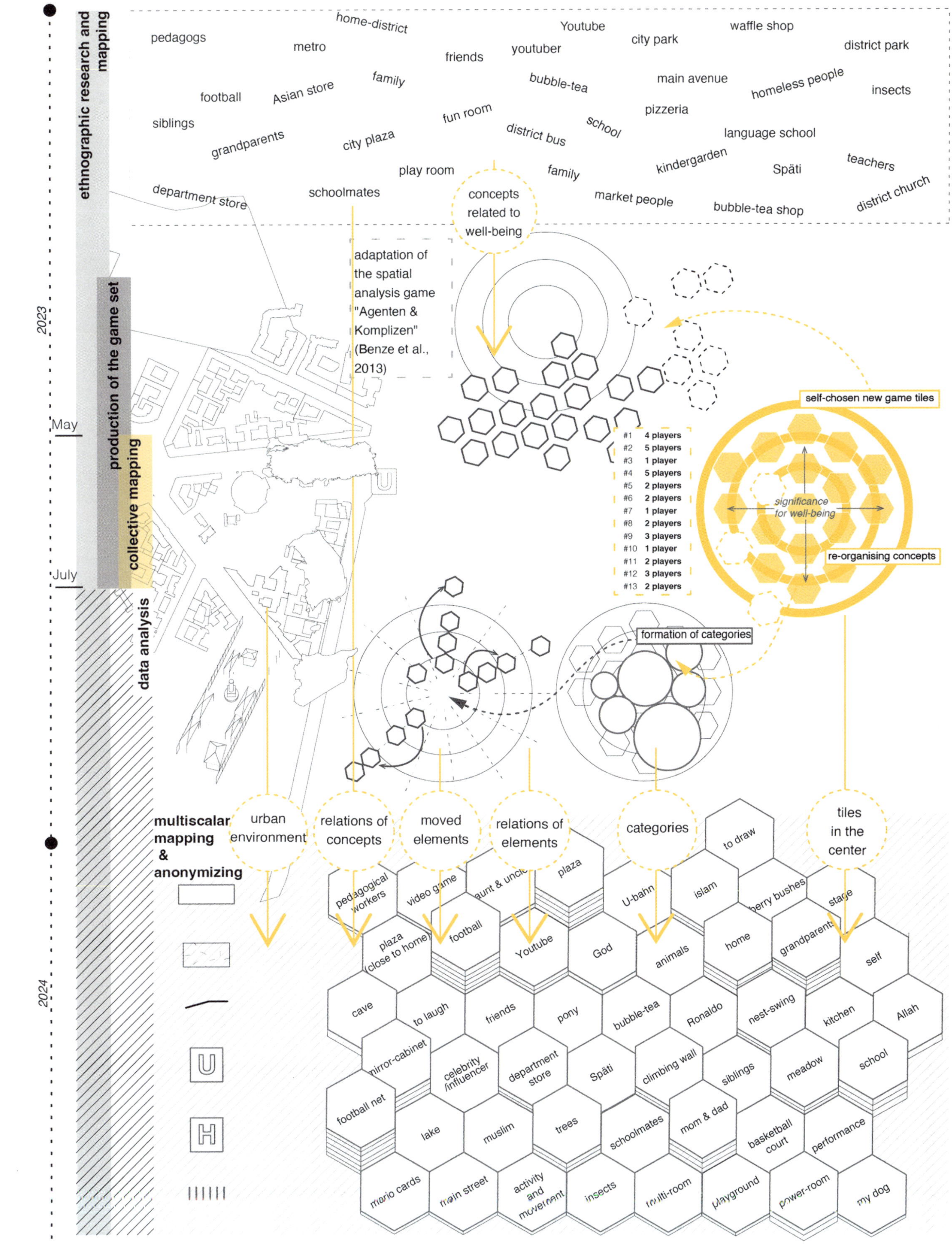

Figure 1 Overview of methodological steps: production of the game set, data collection via collective gamified mapping approach, and visual data analysis (WIKK*I Urban Design Team, 2024)

The first phase was directed by a research team of educational scientists, one of whom was familiar with the children's centre setting from previous studies. During the go-alongs (neighbourhood explorations), data were collected through spontaneous and situational interviews between the participants and the researchers. Observations and interviews were documented using audio recordings, field notes, and observation protocols.

In the second phase of our research, we used the game as a mapping tool for conducting participatory research. We adapted an existing spatial analysis game called 'Agenten & Komplizen' (Benze et al. 2021) to suit our research needs. The game board was restructured into three zones: centre, middle, and periphery. In the context of our mapping tool, 'centre' and 'periphery' function as relational indicators rather than purely spatial categories. The 'centre' denotes elements that participants subjectively prioritise as most important, while the 'periphery' represents aspects of lesser or secondary significance, regardless of physical proximity. By limiting the centre to a single tile, the tool encouraged negotiations among the participating children, emphasising relational dynamics and facilitating deeper discussions about well-being.

The game tiles were developed and produced based on data collected during phase 1 and encompassed people, activities, and places that the children mentioned during the interviews. After the first round we also decided to include empty tiles in the game onto which children could put their own words and terms, ensuring that their subjective perspectives were fully included (see examples of the tiles in Figure 2). During and through the game sessions, we were able to engage the children aged three through twelve actively in the research by creating a playful atmosphere and thereby a pleasant environment. Our approach fostered social interaction, which allowed us to observe how the children made their decisions and set priorities regarding where they feel comfortable or happy and how they experience a sense of agency, security, and self-confidence.

Step 2: Data Collection via Gamified Collective Mapping

The game itself was played in two rounds, with groups of two to five participants. Initially, the participants were asked to create a well-being map on the game board using either the predefined game tiles or new ones created by placing self-chosen terms on blank tiles. Game tiles of high priority for the children were placed in or near the centre, while lower-priority aspects were positioned further outwards on the game board. This arrangement prompted the children to engage in discussions, negotiate priorities, and collectively construct their maps. In this way, a playground—physically close but relationally central—might be placed in the centre, while a distant park—physically far but central in terms of feeling significant—could also occupy a central position.

Thirteen game sessions were conducted with thirty-six children in total (Figure 2). Throughout all phases of the study, emphasis was placed on obtaining the explicit consent of the children. The game sessions were documented using photographs of certain game results in each game session, short field notes, and post-session protocols.

Step 3: Data Analysis through Diagrams and Relational Maps

The data collected during the collective mapping stage were analysed using grounded theory procedures (Corbin and Strauss 1990). Diagrams and graphic elicitation techniques (Bagnoli 2009; Buckley and Waring 2013; Clarke et al. 2022) were used to make spatial patterns and relationships visible, which enabled us to also observe spacetime dynamics within the game. Through this iterative process, we identified seven categories of well-being elements within the topics of game tiles: *people*, referring to individuals or groups explicitly mentioned by the participants; *objects*, which included items, goods, or food specifically written by participants on game cards; *places*, referring

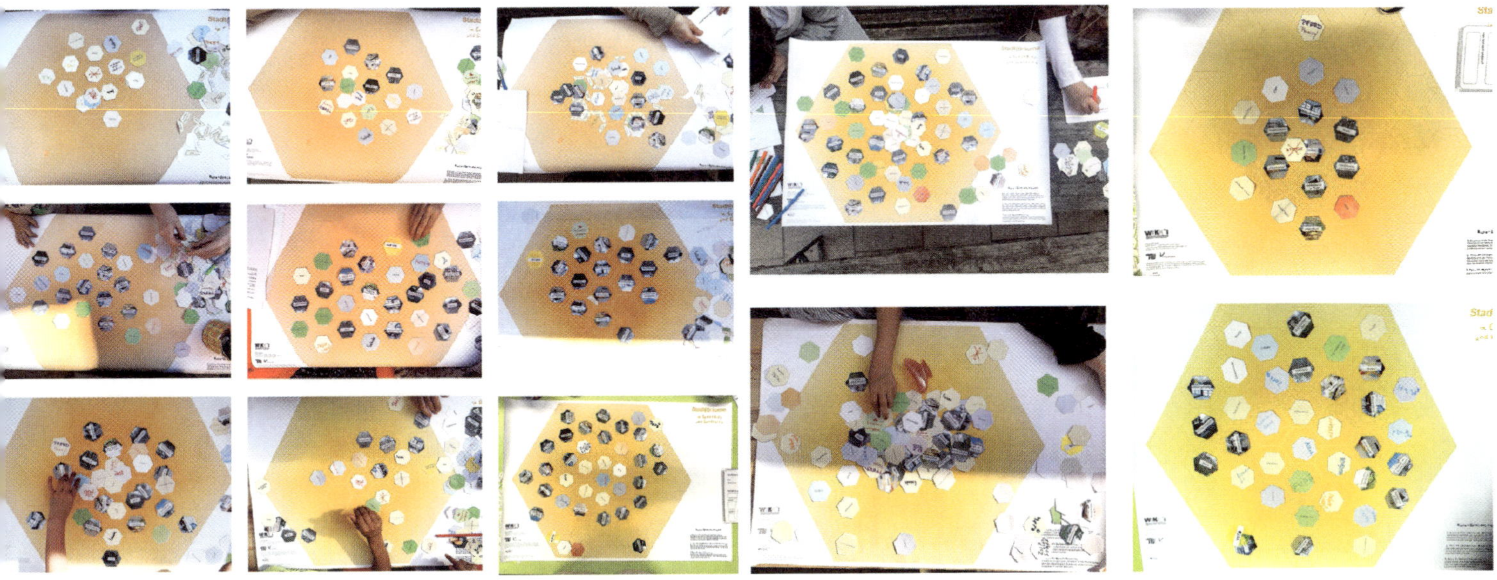

to locations within and outside of the participants' neighbourhood identified as important by the participants; *activities*, including games or sports that participants expressed enjoyment in doing; *religion*, which included religious concepts explicitly mentioned by the participants; *digital media*, referring to digitalised activities and content brought into the conversation by participants; and *nature*, including urban landscapes, flora, and fauna identified by the participants.

As all final game results (Figure 2) were photographed, interim movements and shifts were captured through additional photos during gameplay. These observations were systematically translated into analytical diagrams (Figure 3), in which tiles were grouped into the predefined categories and mapped according to their positions on the game field. Arrows were used to illustrate the flow and transition of tiles, making visible the relational dynamics that Massey (2005) describes as essential to understanding spatial interactions. In Game Session 1 (Figure 3), places such as a 'Central Avenue' and a 'Swimming Pool' initially held central positions but shifted outward as children discussed their mixed experiences at these locations—sometimes reflecting enjoyable moments, other times less favourable memories. In Game Session 4, the shifting positions of significant people, such as 'Teachers' and 'Schoolmates', showcased personal relational dynamics. Gender and cultural differences were also evident, with Game Session 1involving boys making religion more prominent, and Game Session 4 showing exclusively girls making places and personal relationships relevant. However, across all game sessions, religion was not emphasised solely by boys but rather by children from specific cultural backgrounds. The diagrams illustrated also how subjective spatialities and temporalities are negotiated into collective representations while engaging in real-time prioritisations.

A further phase of visual analyses sought to identify patterns in specific aspects of the research focus. For example, Figure 4 consolidates all game tiles placed in the central area of the game boards across all sessions, highlighting the elements considered most significant for the well-being of the participating children. While some tiles are graphically stacked to visually

Figure 2 Photocollage showcasing the final outcomes of thirteen game sessions, highlighting the children's collectively prioritised elements of well-being (WIKK*I Urban Design Team, 2024)

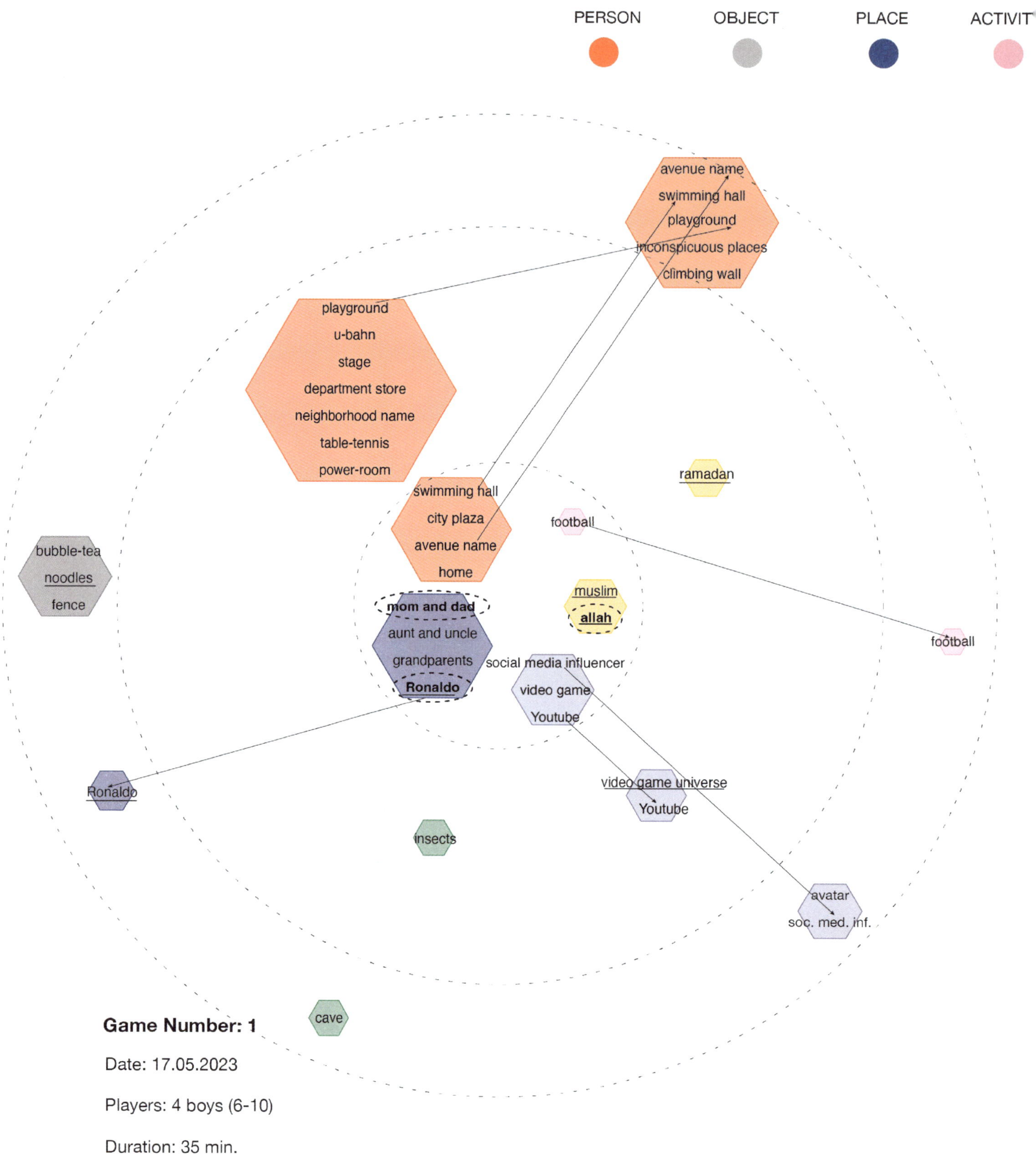

Figure 3 Translation of photos from two different game sessions (left: Game Session 1; right: Game Session 4) into visual diagrams that cluster elements into categories and illustrate shifts in individual elements during the sessions (WIKK*I Urban Design Team, 2024)

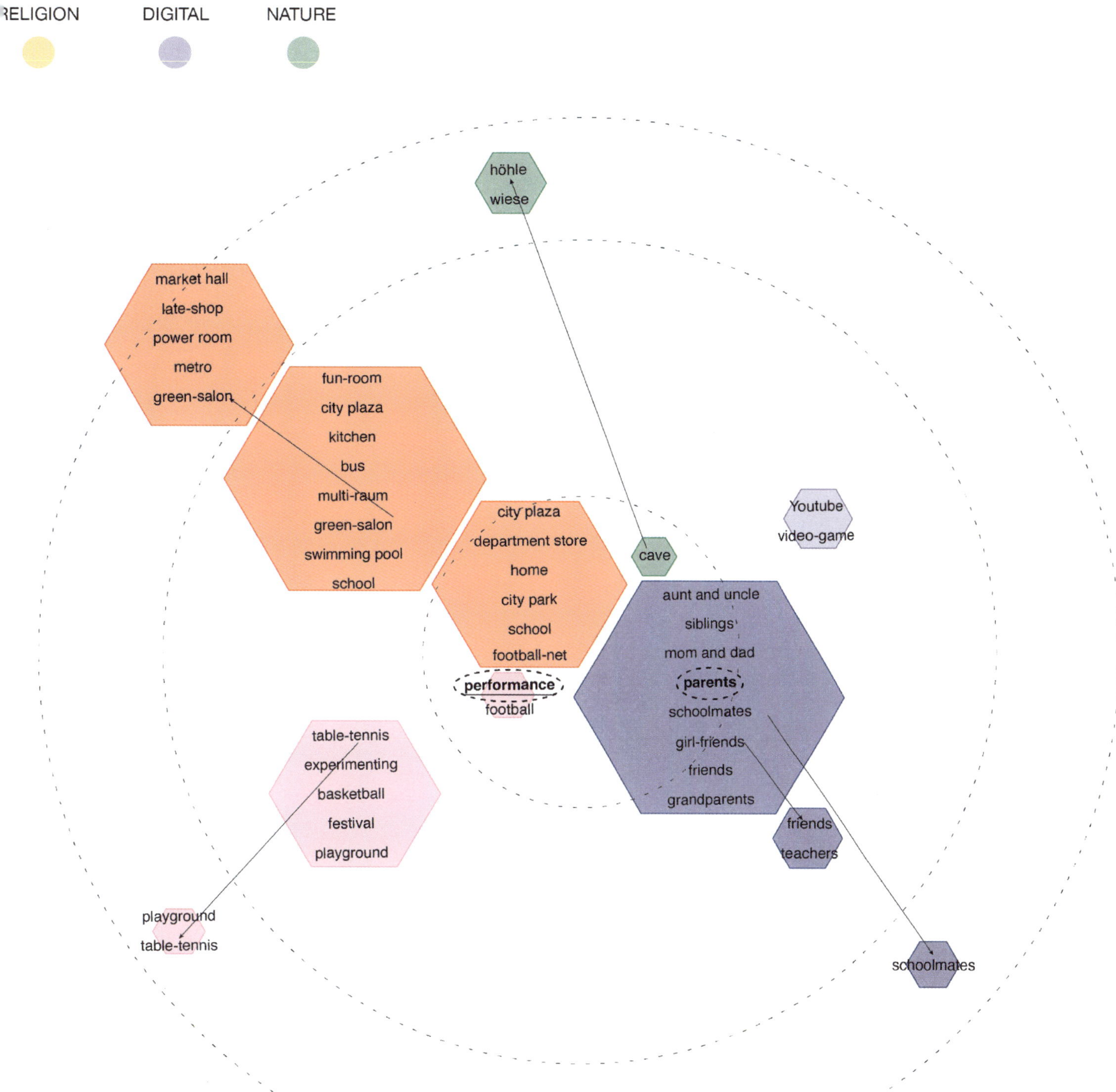

Game Number: 4

Date: 14.06.2023

Players: 5 girls (8-12)

Duration: 30 min.

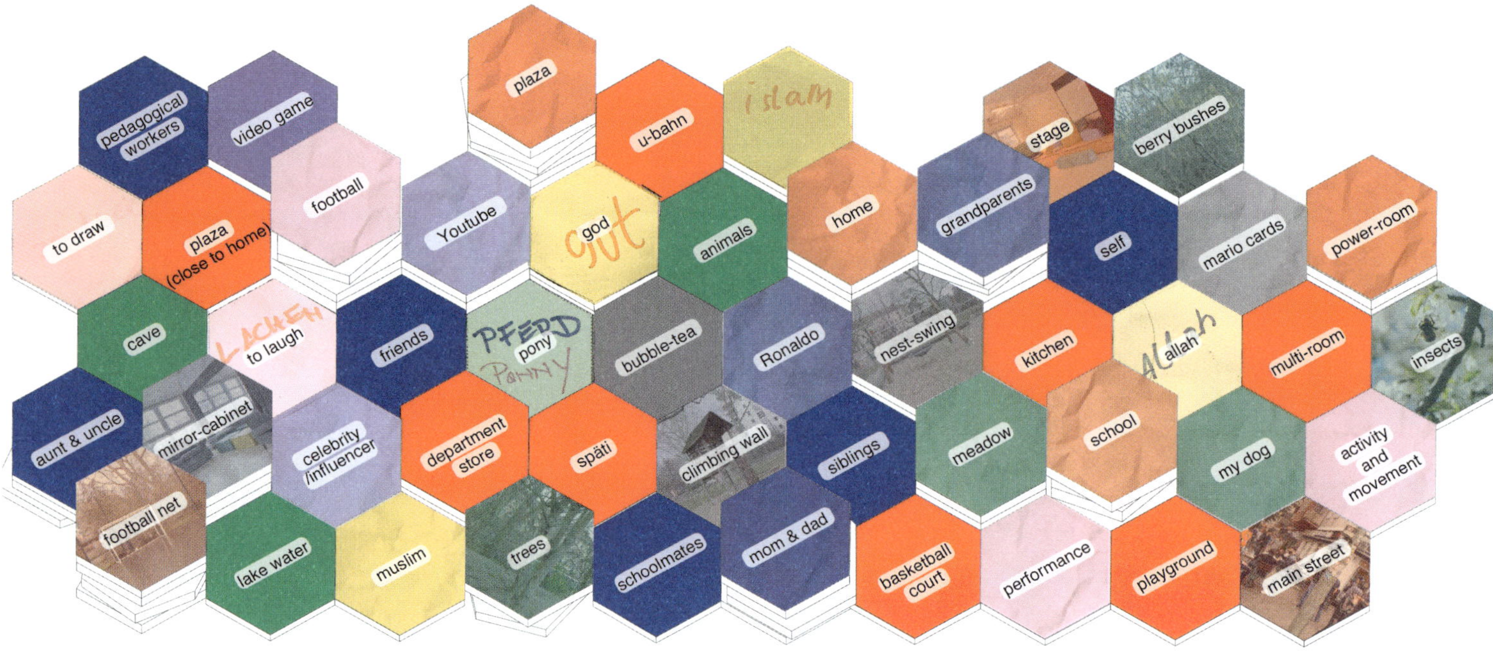

Figure 4: Collage of game tiles placed at the centre of the boards across sessions, representing the elements deemed most central to children's well-being (WIKK*I Urban Design Team, 2024)

represent recurring elements, this is not a quantitative analysis but rather depicts relational priorities of the children's final momentary assessments. The collage does not contain places, objects, people, or activities that shift to lower prioritisation, as indicated in Figures 3 and 5. Such analyses underscore the importance of capturing both the dynamic processes of negotiation and the final outcomes of gamified mapping. The space-time dynamic of gamified mapping benefits from a series of visual analysis products, each offering distinct insights; the dynamic diagrams illustrated the processes and negotiations of prioritisation during gameplay (e.g., Figure 3), whereas the collages consolidated the children's collective final assessments, providing stable representations of socio-spatial priorities at specific moments (e.g., Figure 4).

Further thematic analysis delved deeper into the specific types of places and objects prioritised during the game sessions. Figure 5 highlights how these elements relate to both the children's centre and the broader neighbourhood, revealing additional layers of meaning and spatial significance. The diagram demonstrates the diversity of spaces contributing to children's well-being, spanning both indoor and outdoor environments and objects (e.g., a nest swing or mirror cabinet) and places (e.g., the playground). For example, the children's centre offers a variety of places and objects that were repeatedly mentioned, while neighbourhood spaces, such as school and main streets, complemented these offerings. The temporal dimension was also evident, as certain spaces—such as indoor areas—gained prominence depending on the seasons, reflecting how time and weather shape spatial preferences. By comparing these elements, the analysis emphasised the interconnectedness and variety of spaces that collectively enhance children's well-being, with the children's centre serving as a central and dynamic node. Thus, the thematic diagram captures the relational dynamics of spaces that the children use to construct their well-being.

Step 4: Synthesis of Findings in Multiscalar Mapping

To synthesise and validate the findings, we used multiscalar mapping (Figure 6) as a spatial-analytical instrument that combined insights from phase 1 ethnographic research and phase 2 gamified collective mapping. It allowed us to incorporate heterogeneous data across different scale levels into a single visual representation, including zoomed-in details, text narratives, and symbols (Pelger et al. 2024). It drew on multiple datasets—interviews, observations, game sessions, notes, and analytical graphics.

The content selected for the map highlighted aspects made central by the participating children (Figure 3). For example, *people* frequently mentioned in game sessions, such as family members (mother, father, siblings, close relatives, and grandparents), often dominated the centre of the game board. All individuals placed were located on the map in relation to significant places. Similarly, *objects* positioned at the centre of the game board were also illustrated in their spatial context. *Places* that children love to visit in the neighbourhood were also visualised as important—including several indoor and outdoor spaces and objects, specific shops, and restaurants in the neighbourhood. These visuals were supported with quotations from the children's interviews.

The multiscalar maps also allowed us to illustrate how the children drew spatial references of well-being from their families' countries of origin, where they sometimes spend family vacations. The dynamics of these spaces extended beyond the immediate local level to broader associations, such as the connection between local urban spaces and countries like Turkey or Syria. For example, the local square market was referenced as reminiscent of the family's home country, while the bubble tea shop was more than a favourite refreshment spot—it was associated with South Korea, a country they aspired to visit in the future. Thus, the multiscalar mapping visually

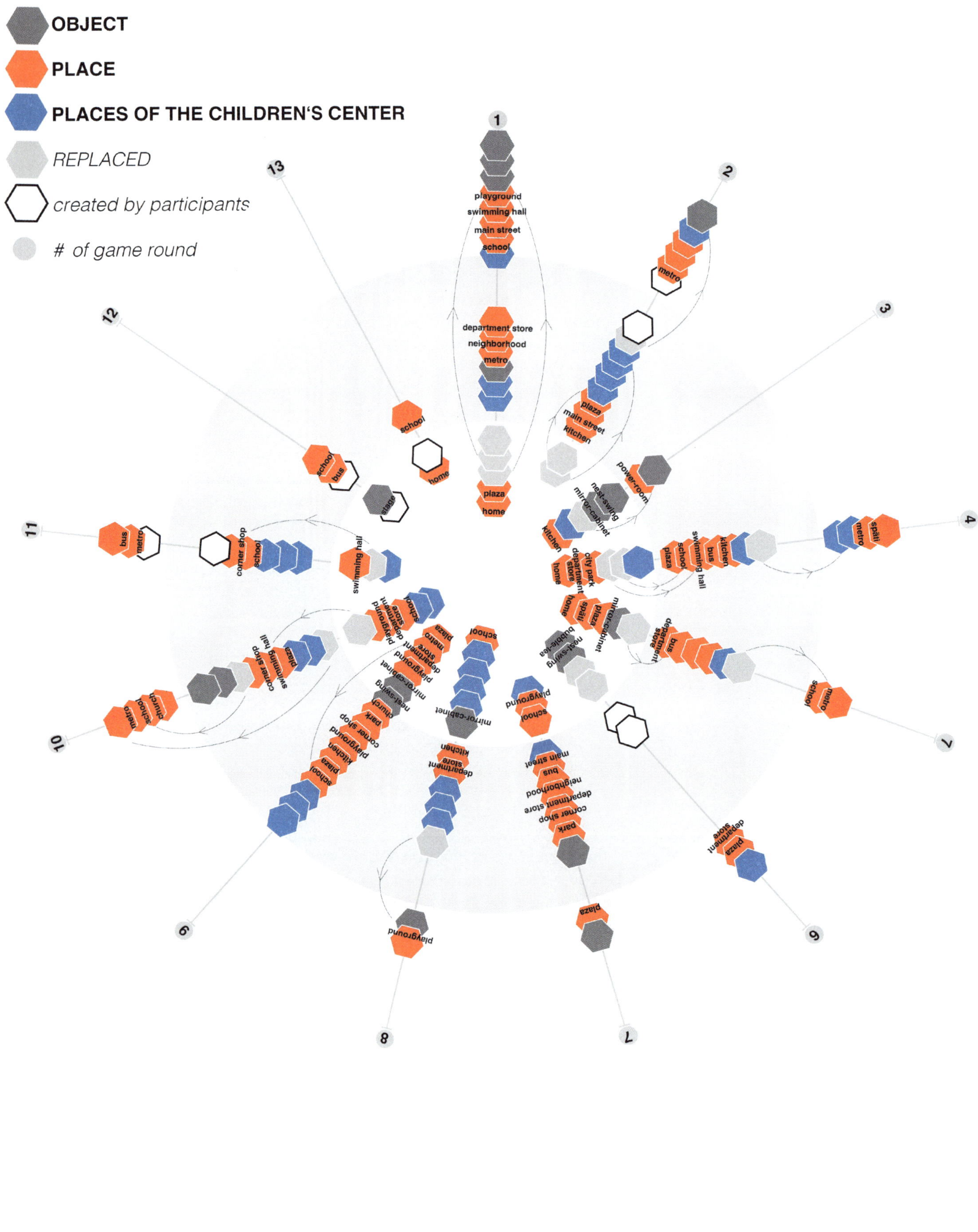

Figure 5 Thematic diagram of places and objects identified by children during game sessions, highlighting the interplay between the children's centre and neighbourhood places and objects for children's well-being. Right: Zoomed-in image of the graphic (WIKK*I Urban Design Team, 2024)

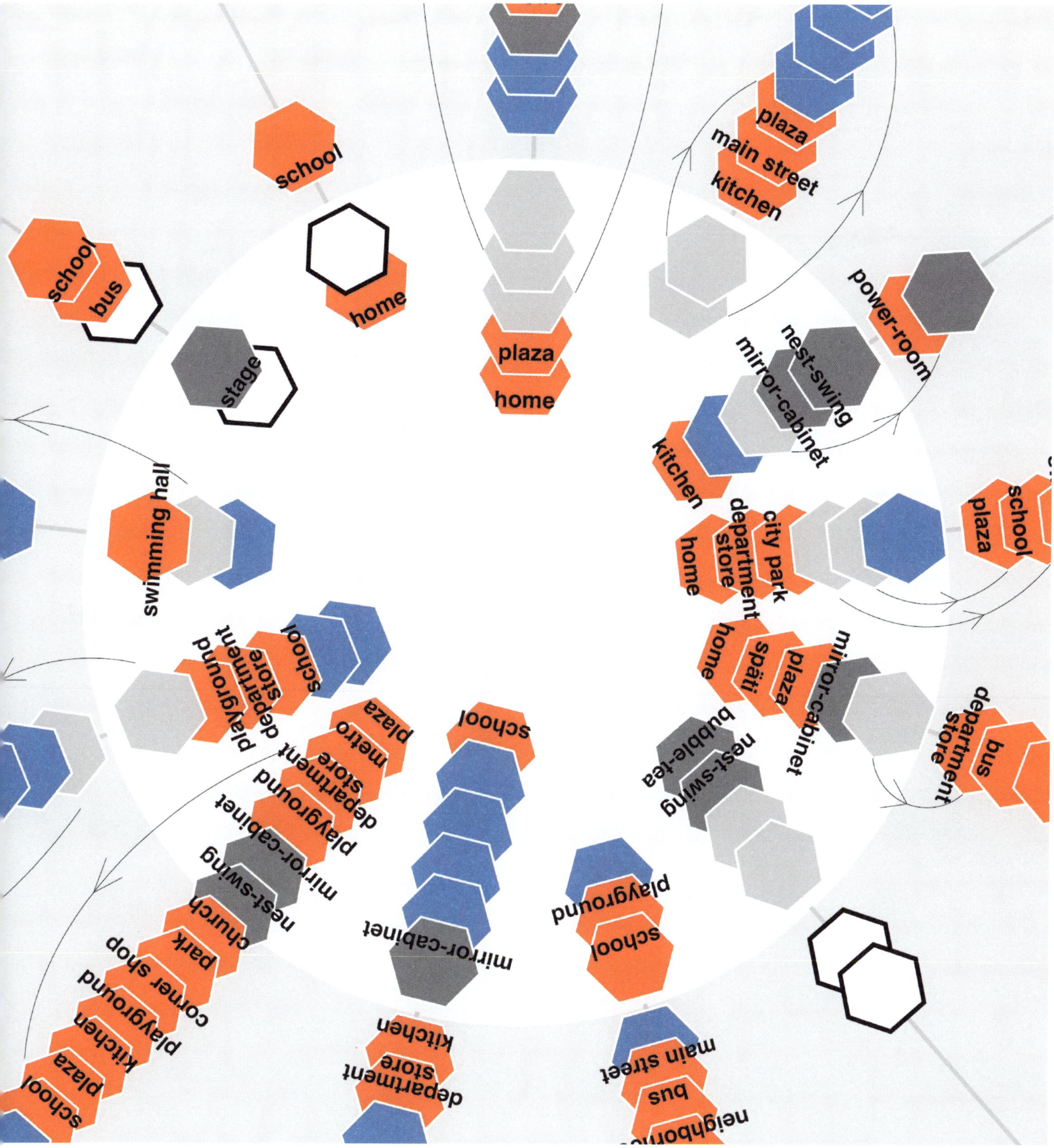
school
school
bus
home
stage
plaza
home
plaza
main street
kitchen
power-room
nest-swing
mirror-cabinet
kitchen
city park
department store
home
school
plaza
swimming hall
home
späti
plaza
mirror-cabinet
bus
department store
bubble-tea
nest-swing
school
playground
department store
school
plaza
metro
department store
playground
mirror-cabinet
nest-swing
church
park
corner shop
playground
kitchen
plaza
school
mirror-cabinet
playground
school
kitchen
department store
main street
bus

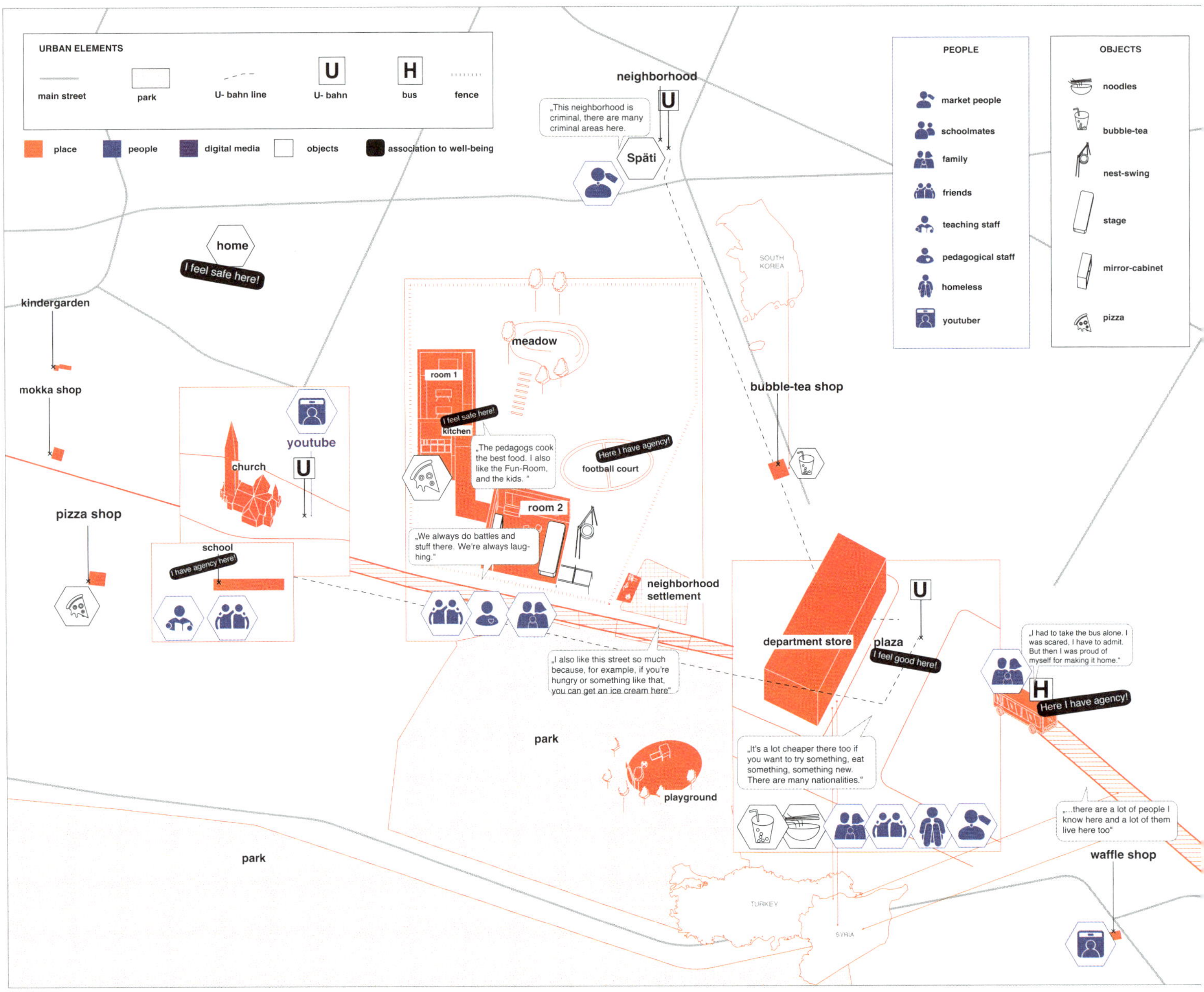

Figure 6: Anonymised multiscalar map illustrating children's well-being through spatial and relational dynamics, balancing ethical data representations (WIKK*I Urban Design Team, 2024)

represented the diversity of coexisting spaces, with varying temporalities also encompassing the present, past, and future spatialities of the children.

These cross-spatial associations were emphasised visually on the map, reinforcing the idea that place is never simply a static backdrop but is rather connected to a larger web of social and cultural relations. This underscores the relational quality of space, where significance is not inherent to a place but is socially constructed and shaped by personal, collective, and historical connections.

Step 5: Anonymising Findings for Dissemination

The process of anonymising the findings for dissemination was undertaken in parallel with the synthesis of multiscalar mapping (Figure 6) and was informed by the legal and ethical considerations outlined at the beginning of this chapter. Anonymising participants and exact place names were a requirement from the start to ensure that the children's centre and its neighbourhood could not be recognised or traced back. This posed a significant challenge, as the names of places, physical structures, and urban qualities are integral to the findings related to urban planning and architecture.

To address this, the study focused on balancing the ethical obligation to protect privacy with the methodological need to preserve the relational dynamics crucial to understanding children's well-being. The challenge lay in finding principles for anonymisation that maintained the visibility of spatial relationalities. This necessitated a nuanced approach: the map discarded quantitative measures, such as distance and size; abandoned specificities of location, such as direction; and replaced singular architectural and urban elements with multiplicities of socio-spatial elements.

The multiscalar mapping itself served as a form of decontextualisation by selectively scaling in and out, which resulted in overlays of map content and zoom-ins. Spatial features, such as roads, plazas, and landmarks, posed a risk of indirect identification. Therefore, we emphasised the abstraction of spatial data without erasing relational insights. Categories such as *places, people, objects*, and *activities* were contextualised within a generalised urban environment. By symbolising common urban elements, such as main arteries, plazas, transportation vehicles, building types, and streets, the anonymised map retained both visual and symbolic relevance while preventing direct traceability.

As Massey (2005) argues, spaces is not merely a neutral backdrop but is actively produced 'in terms of the social relations through which the spaces, and that openness and closure, are constructed; the ever-mobile power-geometries of space-time' (p. 166). This perspective ensured that the map's representation of urban space not only reflects and facilitates an understanding of the relational dynamics between places, people, and objects in children's well-being. It also adhered to Fischer and März's (2024) argument that anonymisation in spatial contexts should prioritise participant protection over the complete erasure of spatial markers. While anonymising spaces entirely might compromise the richness of relational and temporal data, this approach aims at an ethical balance that preserves interpretative depth and relational insights.

As such, the multiscalar map highlighted the methodological compromises inherent in spatial research while aligning with Pelger et al.'s (2024) assertion that ethical spatial mapping requires balancing data protection with the integrity of the research. By discarding quantitative specifics and emphasising relationalities, the resulting map transcended traditional architectural and urban planning map representations to provide an ethical, yet insightful representation of the socio-spatial dynamics explored in this study.

Discussion and Conclusion: Navigating Ethics and Insights in Gamified Mapping of Children's Well-Being

Children's voices and perspectives are often overlooked in discussions about their well-being, yet they have a profound ability to articulate the complex interplay between their environments and experiences. This study explored how gamified collective mapping provides a participatory way to uncover these nuances, transforming individual narratives into collectively negotiated insights about socio-spatial and temporal dynamics. The children's centre provided an ideal setting for implementing this methodology. It combined structured activities and free self-organisation, offering a safe space where children could engage at their own pace. This flexibility allowed for the integration of gamified mapping into their daily routines to ensure both participation and agency. The method's success also lies in its ability to turn subjective data into collectively negotiated insights, highlighting children's active roles as co-constructors of their socio-spatial realities. The gamified approach offers children various means of playful communication, including written, drawn, and spoken expressions, while providing materials such as tiles, pens, and a game board. This variety fosters engagement and enables children to articulate their perspectives in ways that suit their preferences and abilities.

The maps (Figures 2–6) generated during this research revealed not only the children's immediate spatial interactions but also their connections to broader temporalities. While the theoretical framework integrates Massey's space-time concept, empirical insights into temporal dynamics remain limited in our research, representing a key area for further exploration. Nevertheless, the findings demonstrate that children are able to reference past memories, present experiences, and future aspirations within the mapping process, illustrating the space-time nature of their well-being.

In the gamified mapping approach, subjective individual data on well-being are rearranged and negotiated as collective data, creating a rich environment that potentially generates diverse data types, including video, audio, maps, and narratives. However, the full potential of this approach is constrained by ethical research standards and data protection regulations. In this project, we focused primarily on mapping sessions (without video recording), ensuring that the collected data were anonymised and analysed according to these standards and regulations. These considerations significantly shaped not only the setup of the gamified collective mapping but also influenced the visual analysis and visual representation of findings in a multi-scalar map. The multi-scalar map itself represents a methodological compromise, as it abstracts sensitive details while maintaining symbolic connections to urban contexts. This approach reframes anonymisation not as a limitation but as a reflective practice that enhances ethical accountability and methodological transparency. By embedding ethical considerations into the mapping process, we want to contribute to ongoing debates about anonymisation in spatial mapping and research. In doing so, we acknowledge the challenge of conveying findings that reflect the multiscalar and socio-spatial dimensions relevant to children's well-being—both within the academic community and professional planning and design practice—, while preserving the depth and richness of the children's experiences.

Despite these challenges, we found that the synthesis maps and visual analysis graphics proved highly effective in facilitating discussions within the research community and among practitioners. Their presentation at conferences sparked valuable dialogues about specific types of multiscalar spaces that influence children's well-being. The synthesis maps serve as critical tools for academic discourse by illustrating complex spatial–temporal dynamics, while the game itself offers a direct, engaging medium for practitioners to explore children's well-being *in situ*. In contrast, the game showed distinct strengths in direct application. Presented and play-tested in two workshops with practitioners, the game elicited enthusiastic feedback, with participants recognising its potential as a practical tool for engaging with children. Practitioners suggested using the game in their everyday work to better understand what matters most to children in regard to their well-being. To facilitate this, the game board has been made available as an insert in a brochure, and all game templates—including tiles and a printable poster of the game board—

are accessible online as downloadable resources. This practical availability ensures the method's broader applicability and encourages its use in diverse contexts, bridging research and practice.

Reflecting on the potentials of gamified collective mapping, the approach effectively encouraged playful communication, discussion, and collective decision-making. These mechanics facilitated the integration of children's perspectives, enabling them to actively engage in the co-creation of socio-spatial narratives about their well-being.

Funding Statement

This research was funded by the Federal Ministry of Education and Research of Germany (BMBF) research programme: 'Overcoming Educational Barriers: Learning Environments, Educational Success, and Social Participation', 2022–2024, Project name: 'Well-being in Socio-Spatial Contexts: Intersectional Perspectives on Children's Experiences at Non-School Learning Sites (WIKK*I)'

Image Credits

Figure 1, 3, 4, 5, 6: WIKK*I Urban Design Team
Figure 2: Participants and WIKK*I Urban Design Team

References

Bagnoli, Anna. 2009. 'Beyond the Standard Interview: The Use of Graphic Elicitation and Arts-Based Methods'. *Qualitative Research* 9 (5): pp. 547–570. https://doi.org/10.1177/1468794109343625.

Ben-Arieh, Asher. 2006. *Measuring and Monitoring the Well-Being of Young Children around the World*. France: UNESCO.

Benze, Andrea, Christina Jimenez Mattsson, and Urs Walter. 2021. 'Games as Urban Agora. An Analysis of Games as Participatory Research, Co-Design, and Educational Tools in Urban Planning'. In *The Routledge Companion to Games in Architecture and Urban Planning: Tools for Resign, Teaching, and research*, edited by Marta Brkovic Dodig and Linda N. Groat, pp. 221–233. New York: Routledge.

Brkovic Dodig, Marta, and Linda N. Groat, eds. 2021. *The Routledge Companion to Games in Architecture and Urban Planning: Tools for Design, Teaching, and Research*. New York: Routledge.

Buckley, Charles A., and Michael J. Waring. 2013. 'Using Diagrams to Support the Research Process: Examples from Grounded Theory'. *Qualitative Research* 13 (2): pp. 148–172. https://doi.org/10.1177/14687941124 72280.

Castillo Ulloa, Ignacio, Anna J. Heinrich, Angela Million, and Jona Schwerer. 2024. *The Evolution of Young People's Spatial Knowledge*. Abingdon: Routledge.

Chassé, Karl A., Margherita Zander, and Konstanze Rasch. 2003. 'Sozialpädagogische Konsequenzen'. In *Meine Familie ist arm*, edited by Karl A. Chassé, Margherita Zander, and Konstanze Rasch, pp. 322–343. Wiesbaden: VS Verlag für Sozialwissenschaften. https://doi. org/10.1007/978-3-322-93453-6_6.

Christmann, Bernd, Daniel Schwerdt, and Martin Wazlawik. 2018. 'Umgang mit ethischen Fragen in der Forschung zu sexualisierter Gewalt an Kindern und Jugendlichen: Erfahrungen und Befunde im Rahmen des Forschungsnetzwerks "Sexuelle Gewalt in pädagogischen Kontexten"'. *Soziale Passagen* 10 (1): pp. 121–133. https://doi. org/10.1007/s12592-018-0286-5.

Clark, Alison 2017. *Listening to Young Children: A Guide to Understanding and Using the Mosaic Approach*. Expanded 3rd ed. London: Jessica Kingsley Publishers.

Clarke, Adele E., Rachel Washburn, and Carrie Friese, eds. 2022. *Situational Analysis in Practice: Mapping Relationalities Across Disciplines*. New York, NY: Routledge. https://doi.org/10.4324/9781003035923.

Corbin, Juliet M., and Anselm Strauss. 1990. 'Grounded Theory Research: Procedures, Canons, and Evaluative Criteria'. *Qualitative Sociology* 13 (1): pp. 3–21. https://doi.org/10.1007/BF00988593.

Cortés-Morales, Susana, Louise Holt, Jenny Acevedo-Rincón, Stuart Aitken, Danielle Ekman Ladru, Tanja Joelsson, Peter Kraftl, Lesley Murray, and Gabriela Tebet. 2022. 'Children Living in Pandemic Times: A Geographical, Transnational and Situated View'. *Children's Geographies* 20 (4), pp. 381–391. https://doi.org/10.1080/14733285.20 21.1928603.

DeWalt, Katleen M., and Billie R. DeWalt. 1998. 'Participant Observation'. In *Handbook of Methods in Cultural Anthropology*, edited by H. Russel Bernard, pp. 259–300. Walnut Creek: AltaMira Press.

Fattore, Tobia, Jan Mason, and Elizabeth Watson. 2017. *Children's Understandings of Well-Being: Towards a Child Standpoint*. 1st ed. Children's Well-Being: Indicators and Research 14. Dordrecht: Springer Netherlands.

Fischer, Lisa, and Stella März. 2025, forthcoming. 'Legal Protection to Privacy and Consent in Research with Children as an Ethical Inequality Problem'. In *Qualitative Fieldwork with Children. Context and Participation in Child Well-Being Research across Nations*, edited by Lise Mogensen, Susann Fegter, Lisa Fischer, Jan Mason, and Tobias Fattore. Bristol: Bristol University Press.

International Child Development Centre, ed. 2016. *Fairness for Children: A League Table of Inequality in Child Well-Being in Rich Countries*. Florence: UNICEF Office of Research – Innocenti.

Kessl, Fabian, and Christian Reutlinger, eds. 2022. *Sozialraum: Eine elementare Einführung*. Wiesbaden: Springer VS. https://doi. org/10.1007/978-3-658-29210-2.

Kusenbach, Margarethe. 2008. 'Mitgehen als Methode. Der "Go-Along" in der phänomenologischen Forschungspraxis'. In *Phänomenologie und Soziologie: Theoretische Positionen, aktuelle Problemfelder und empirische Umsetzungen*, edited by Jürgen Raab, Michaela Pfadenhauer, Peter Stegmaier, Joachim Dreher, and Bernt Schnettler, pp. 349–358. Vallendar: VS Verlag für Sozialwissenschaften. https://doi.org/10.1007/978-3-531-91037-6_31.

Massey, Doreen. 2005. *For Space*. Los Angeles: SAGE.

Million, Angela. 2021. '"No One Listens to Us..." COVID-19 and Its Socio-Spatial Impact on Children and Young People in Germany'. *Children's Geographies*: 20 (4), pp. 469–477. https://doi.org/10.1080/14733285.2 021.1908520.

Million, Angela, Katrin Schamun, and Susann Fegter. 2024. 'Understanding Well-Being Through Children's Eyes: Lessons for Shaping the Built Environment'. *Urban Planning* 9, Article 8618. https://doi.org/ 10.17645/up.8618.

Pelger, Dagmar, Emilie Kelling, and Jörg Stollmann. 2024. 'Multiskalares Mapping'. In *Handbook of Qualitative and Visual Methods in Spatial Research*, edited by Anna J. Heinrich, Severin Marguin, Angela Million, and Jörg Stollmann, pp. 247–262. Bielefeld: transcript Verlag.

Reutlinger, Christian, Caroline Fritsche, and Eva Lingg, eds. 2010. *Raumwissenschaftliche Basics. Eine Einführung für die Soziale Arbeit*. Bd. 7. Sozialraumforschung und Sozialraumarbeit. Wiesbaden: VS Verlag für Sozialwissenschaften.

World Vision E.V. 2013. *Kinder in Deutschland. 3. World Vision Kinderstudie*, edited by Sabine Andresen and Klaus Hurrelmann. Weinheim/Basel: Beltz.

World Vision e. V. 2018. *Kinder in Deutschland. 4. World Vision Kinderstudie*, edited by Sabine Andresen and Sascha Neumann. Weinheim/Basel: Beltz.

Historical-Network Mapping

Qusay Amer

Urban Transformations, Migration, and Refugee Movements in Amman

Introduction: Every City is a Migration City

'Every city is a migration city' is a provoking statement that challenges political and academic discourses. A review of migration studies revealed that cities are often seen as the main hubs or portals for migration (Pisarevskaya and Scholten 2022, 249). In our research, we try to address migration as a crucial feature of urban spaces—whether internal or international, historical or contemporary. Migration is fundamental to the identity, diversity, and development of every city, as it shapes the economy, culture, and social dynamics.

Migration movements and forced displacement have been significantly shaping our cities for centuries. The polarisation among various political actors in governing migration and displacement is a result of the complex relationship between political ideologies and economic interests. While these themes have been intensively researched in different disciplines, the intersection of migration and refugee studies with urban planning is still an emerging field. Embedded in a spatial theory research consortium, the Collaborative Research Centre's *Re-Figuration of Spaces* (CRC 1265), our research project *Architectures of Asylum* focuses on relational space theory by asking which 'glocal' governance constellations emerge in the management of refugee migration in the contexts of two cities—Amman, Jordan, which this article will discuss, and Lagos, Nigeria, explored in detail by my colleague and teammate Francesca Ceola in this volume—and how these constellations influence the physical-material constitution of space through asylum architectures. The project expands the spectrum of the architectures of asylum to be studied by engaging with typologies beyond refugee camps and emergency shelters. Thus, the assumption that global refugee migration is a critical juncture becomes an opportunity to rethink urban crisis management based on the migration embedded in it. Furthermore, this approach also allows for the consideration of a wider group of refugees (internally displaced persons) and migrants, between whom it is increasingly difficult to draw clear distinctions. The research project utilises various mapping methodologies, which are discussed in these two chapters. The concrete planning negotiations and the physical-material interventions are mapped and analysed through site visits (Tabačková 2021). These interventions and local spatial appropriations initiated by refugees and/or other local residents are recorded graphically. The role of the different actors in implementing the projects is analysed along with translocal knowledge and is presented through methods of multiscalar mapping (Pelger et al. 2021).

The Social Relevance of Migration

Following the understanding of space as relational (Löw 2016), Gunter Weidenhaus (as summarised by Löw 2016, xx) 'speaks of "social space-time" when the constitution of space and that of time logically belong together. The simple fact that an event takes place sometime and somewhere without a systematic interrelation is no reason to assume a coherence of space and time.' We understand from this that various ways of understanding space(s) and time(s) are interconnected in patterns that have social relevance (Weidenhaus 2023). Since migration and refugee movements take place over time and affect the spaces where they happen, our project emphasises their social relevance. This impacts not only the spatial and temporal determinations of people (Weidenhaus 2023) but also the place—in our context, 'the urban'.

In our research, we grapple with the challenge of following a continuously displaced population and the shifting spacing practices. Methodologically, this translates into inscribing movement within the research design and experimenting with mapping methodologies that capture space with time, and vice versa. Considering maps as subjective powerful descriptions, possibly deep ones (Ng-Chan 2015), we are reminded of Doreen Massey's (2005, 9) suggestion that space is 'the product of interrelations ... always under construction' and 'a simultaneity of stories-so-far'. Thus, we interpret that maping —that is, retracing, following, analysing, and representing—the spatialities of

A brief history of Amman - Refugee Movements

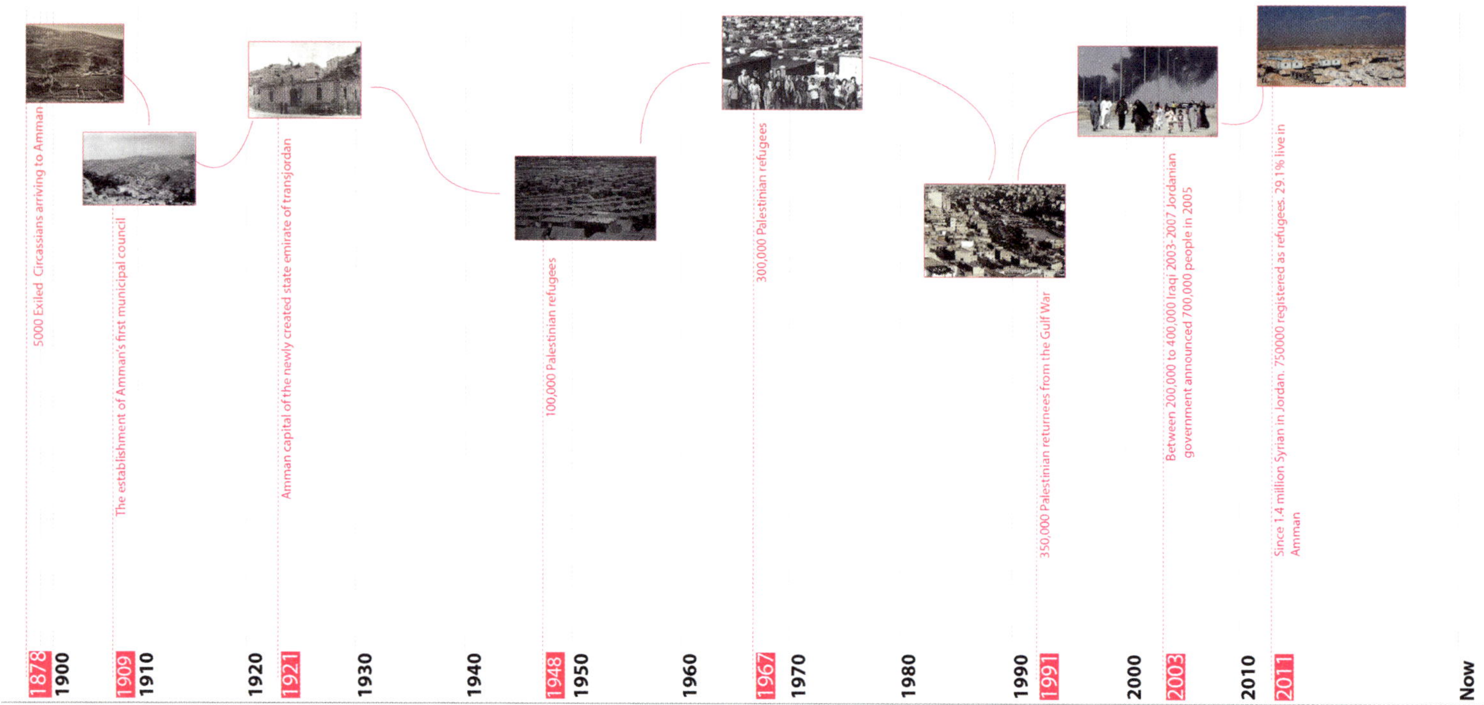

A brief history of Amman - Networks

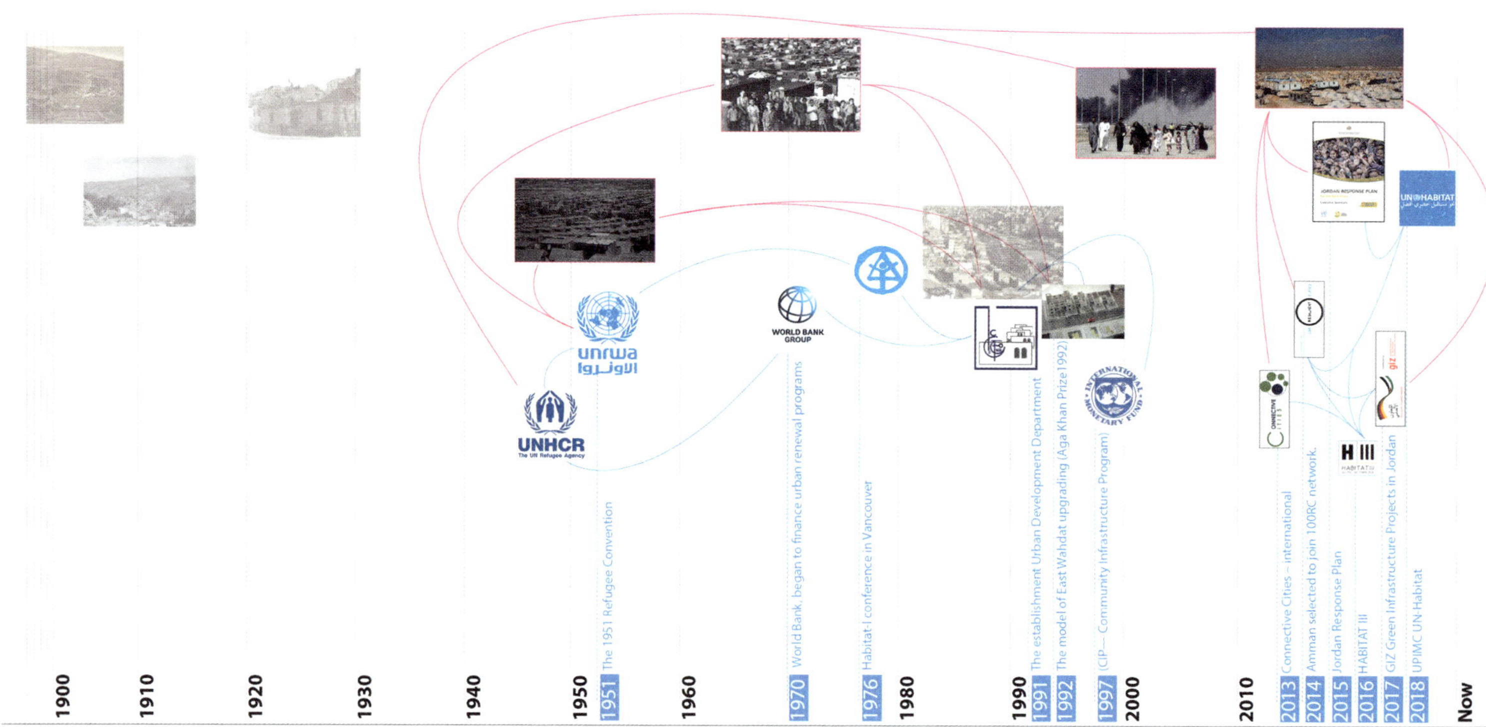

Figure 1 Refugee movement towards Amman (Qusay Amer, 2024)
Figure 2 Historical regulation of migration towards Amman through city networks (Qusay Amer, 2024)

urban refugees' emplacement in the city should not be considered chronological subsequent phases of one process, since they are formed by intricate multilayered experiences and sociopolitical dynamics that challenge linearity. Instead, they should be considered diachronic and a different unfolding of similar processes every time—at once singular in their specificities and collective in traits.

This chapter addresses the methodological approach to understanding how migration is intertwined with Amman's urban development, using Weidenhaus's (2023) concept of 'social space-time' as an integrated framework for understanding these interconnected aspects. In addition, I illustrate the different mapping methodologies used in order to understand the complexities of the spatial and temporal relationships.

Amman as a Refuge

Amman, the capital of Jordan, known as the 'city of immigrants' (Al Hamarneh 2002) and the 'city of many hats' (Daher 2011), is sometimes more critically described as being 'too inclusive' (Ghanem 2022; Shami 2007). Amman is a living example of a city shaped by different migration and refugee movements. Looking at the components that comprise its history reveals how the emergence and development of migration policy within the city is entangled in urban planning and architecture. My research in Amman focuses on the emerging translocal networks and their influence on local planning policies and governance approaches in dealing with migration, particularly since the Syrian refugees' influx in 2012. This includes looking at the planning and production of physical interventions in different urban spaces and the participation of all actors involved in the process.

The Jordanian government claims that the country hosts 1.4 million Syrian refugees, of which only 670,000 are registered (United Nations Human Settlements Programme [UN-Habitat] 2022). According to Al Tal and Ghanem (2019), approximately 83 per cent of all refugees, who are identified as 'non-camp refugees', have settled in the urban areas of Amman and the northern governorates of Jordan instead of the designated camps; meanwhile, the remaining 17 per cent live in refugee camps (Ministry of Planning and International Cooperation [MoPIC] 2015). My research in Amman points to the multilayered nature of the policies, which are linked to geopolitical agendas, which, in turn, imply a network of relationships among diverse actors operating within and beyond a formal institutional framework. The research shows that these interactions are influenced by historical legacies, legal structures, and informal dynamics. Regarding the social relevance of migration, space-time relations are crucial in this case, especially as I look at the city boundaries, infrastructure, urban agency of locals and refugees, and the neoliberal policies that push refugees into protracted displacement (Brun and Fábos 2015)—a sort of limbo without sustainable perspectives, where at every displacement, the protagonists are forced to start from zero again and again. The next paragraphs will further explore these space-time entanglements.

How Did the City Emerge through Time?

Whoever discusses Amman will have to mention the terms *refugee* and *migrant*. Our research aims to analyse the different refugee migration movements and the governance approaches that had been implemented in the city over decades and compare them to understand 1) how this influences the urban transformations of cities like Amman, 2) how translocal networks impact local planning policies in dealing with migration, 3) which spatial conflicts and negotiation processes shape the places of refugee care in urban contexts, and 4) what the concrete spatial manifestations emerging as architectural or infrastructural products from spatial conflicts look like.

The use of the mapping methodologies of qualitative spatial research was crucial in exploring space, collecting stories, observing, and visualising (Heinrich et al. 2024). To analyse the interconnection between space and time,

I decided to use historical and network mapping methodologies. Historical mapping specifically helped clarify various spatial relationships, the power dynamics within the specific contexts, and patterns that are historically part of the space production process. Network mapping served as a tool to develop a multilevel analysis and create visuals that synthesise and represent the above and the findings in unified objects—a glimpse of which can be seen in the images presented in this chapter.

Historical Mapping

I used historical mapping to understand how migration, displacement, and refuge have changed the urban fabric of Amman. This was based on a literature review and interviews with scholars who have also researched the topic. By comparing maps and specific historical events across different time periods, with a focus on migration patterns, I was able to see how the urban area has changed over time. A central challenge of the historical mapping was sequentiality. Despite the dates that define the axis of the illustration (Figure 1), the thematic passages (political, humanitarian, economic, etc.) challenge linearity, as they do not adhere to a chronological sequence but rather to a complex network of simultaneous realities. Migration in the context of displacement and refuge is often not a straightforward journey, as refugees' journeys often jump back and forth in time and space. They may physically leave their homes, but mentally, they remain at the point of departure. In fact, in most of the contexts I have studied, refugees are kept in a limbo that looks like an empty circle. This is why I decided to analyse and present this historical context not in a strictly linear way but rather as a multi-layered one that contains many contradictions and circularities.

Pelger's (2022) work refers to Lucius Burckhardt's description of planning and design as a threefold interaction of politics, the environment, and mankind. I adjusted those interactions and looked at the history of the city considering three factors that influence the development of the urban fabric: 1) mankind, which involves the joint movement of refugees and migrants over the last decades, 2) politics, which includes the different ways in which administrations work, and 3) the environment, which involves the collective use of the space that leads to urban growth. Figure 1 depicts the timeline with ten-year gaps and the three factors arranged vertically to allow for the possibility of these events overlapping.

Mapping Migration and Urban Interventions in Amman

To understand Amman's urban transformation throughout the last century, I started by mapping refugee movements (Figure 1). After a period in which there were no permanent populations, the first lasting settlement appeared in Amman in 1878, consisting of the exiled Circassian tribes who escaped religious persecution from Russia (Al Hamarneh 2002; Chatty 2010). In 1921, with the fall of the Ottoman Empire, the city became the capital of the newly created state emirate of Transjordan and later became the capital of the Hashemite Kingdom of Jordan. At that time, merchants from Syria and Palestine moved to the town, thereafter to be joined by Armenian immigrants, who escaped the massacre of their people during the First World War (Al Hamarneh 2002; Munif 2018; Hamed-Troyansky 2017). The year 1948 marked a critical date in the region, as more than 100,000 Palestinians arrived in Jordan, and thousands of refugees settled in the new capital as a result. At the time, Jordan expanded its hold over the West Bank of Palestine, all the way to Jerusalem. This was followed by the arrival of 300,000 Palestinian refugees in the country in 1967, when thousands also settled in the capital (Ababsa 2012; Al Hamarneh 2002), as the State of Israel occupied Jerusalem and the West Bank. This effectively expanded Israel's hold over historic Palestine. The Palestinian influx into Jordan continued long afterward. Thus, in 1991, 300,000–350,000 Jordanians, Jordanians with Palestinian origin, and Palestinians were expelled from the Gulf countries after the Gulf War, and they either went back to or were resettled in Jordan (Ababsa 2012). The Gulf War created a major displacement of Iraqi populations. However, the number of Iraqi refugees is controversial. Between 2003 and 2007, and after the US American invasion of Iraq, between 200,000 and 400,000 Iraqis fled to Jordan, while the Jordanian government later claimed that they hosted 700,000 Iraqi refugees in 2005 to justify its requests for international aid, but this figure has been revised significantly downwards (Ababsa 2012). Since 2012, and after the Syrian uprising and the conflict in the country, Jordan received many Syrian refugees. Approximately 29 per cent of the mentioned number reside in Amman, while the rest are spread throughout the country's northern governorates, such as Mafraq and Irbid (Al Tal and Ghanem 2019; Department of Statistics 2015; UN-Habitat 2022).

To understand the refugee movement's influence on Amman's urban development, I added a second factor (Figure 2), which retraces the historical regulation of migration through the city networks of influence and foreign investment. In response to this and the arrival of Palestinian refugees in Amman, four camps were created, which grew massively after 1967. During the 1970s, the World Bank began to finance urban renewal programmes that were strongly inspired by the theories developed by urban planner John F. C. Turner in Peru in the 1960s. Ababsa (2012) explains that, during the 1970s, the World Bank became a main actor financing the urban renewal programme in Jordan, especially in Amman. In this context, in the 1980s, a new agency was created inside the Greater Municipality of Amman's (GAM) Urban Development Department. From 1980 to 1997, Jordan became the first Arab country to implement the new developmentalist ideology first promoted by the World Bank in Latin America and Asia. According to decolonisation scholars, such as Silvia Federici, this ideology resulted in more exploitation and marginalisation of various populations, for instance, through increased debts and privatisation. In 1997, a new policy was imposed by the International Monetary Fund—a huge urban infrastructure programme for the community, both in camps and informal areas (Ababsa 2012). Since the arrival of the Syrian refugees, the city of Amman developed several response plans and strategies, one of which involved becoming an active member of different international networks, such as Connective Cities, the 100 Resilient Cities (100RC) programme, and the C40 network. The last factor I developed was the collective use of the space that leads to an urban growth or 'environment', referring back to Pelger (2022) (Figure 3). Comparing the evolution of city boundaries over time with the migration factor points to the latter as a major contributor to the expansion of city boundaries. Figure 4 shows an overview of the historical mapping and provides a sense of the interwovenness of the migration history of the city. Through the process, I learned about not only the specificities of each case but also the impacts affecting the future of the refugees or even their past.

Network Mapping

Network mapping as a methodology is driven by actor-network theory and is used to visualise the complex dynamics and flows between different entities. As mentioned previously, we used this methodology to analyse and present all of the different actors involved in the production of spaces, their relations, negotiations, conflicts, and strategies. The spatial implications and overlooked connections are shown through this mapping. As previously discussed, space is considered a process and not a fixed entity, so we can see the changes over decades while looking at the actors' network. Figure 5 indicates the various policies and agendas that have occurred over the last decade, addressing several actors. However, it lacks the ability to address all actors equally and to provide a more nuanced understanding of the spatial dynamics. Furthermore, it lacks the simultaneity of the actions that demonstrate the negotiations and spatial events that occur at the same time. All actors have different resources that influence both the projects on the ground and the

A brief history of Amman - Urban Growth

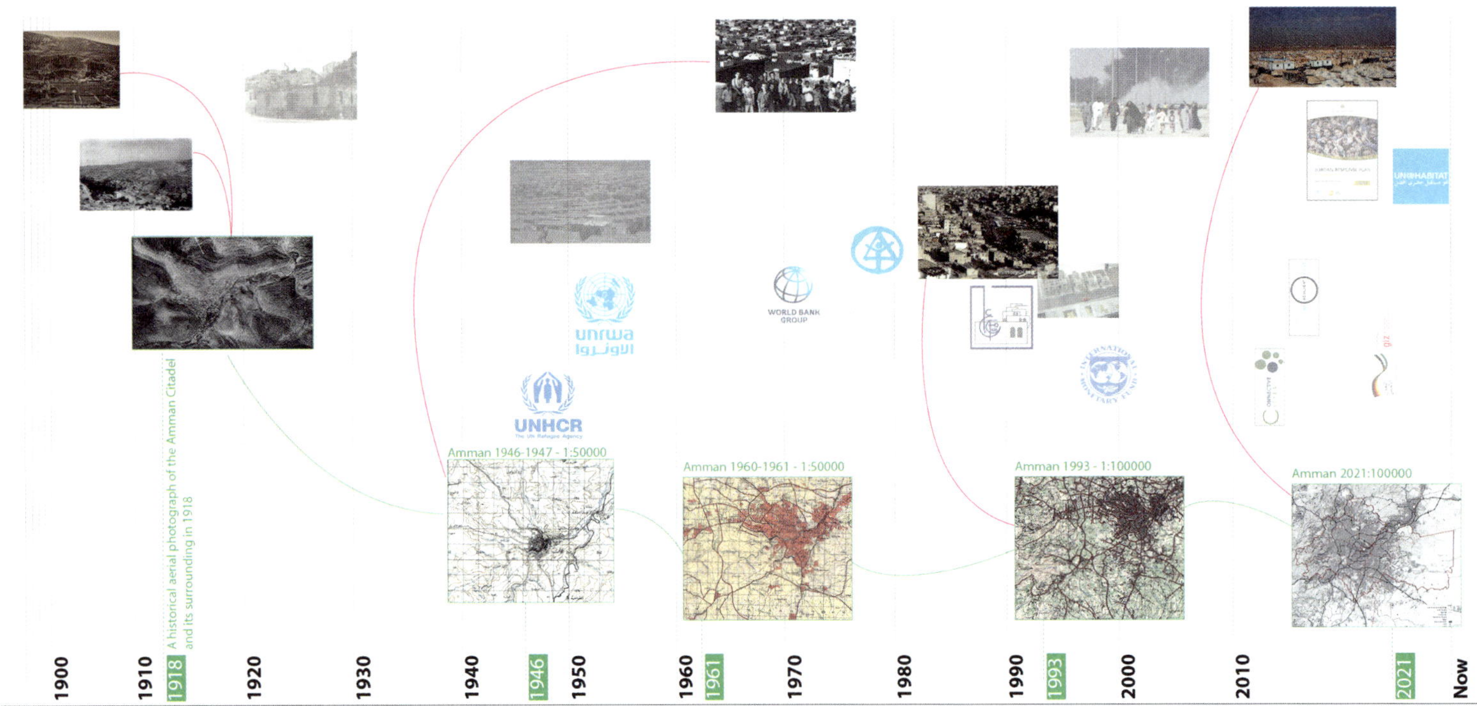

A brief history of Amman - Overview

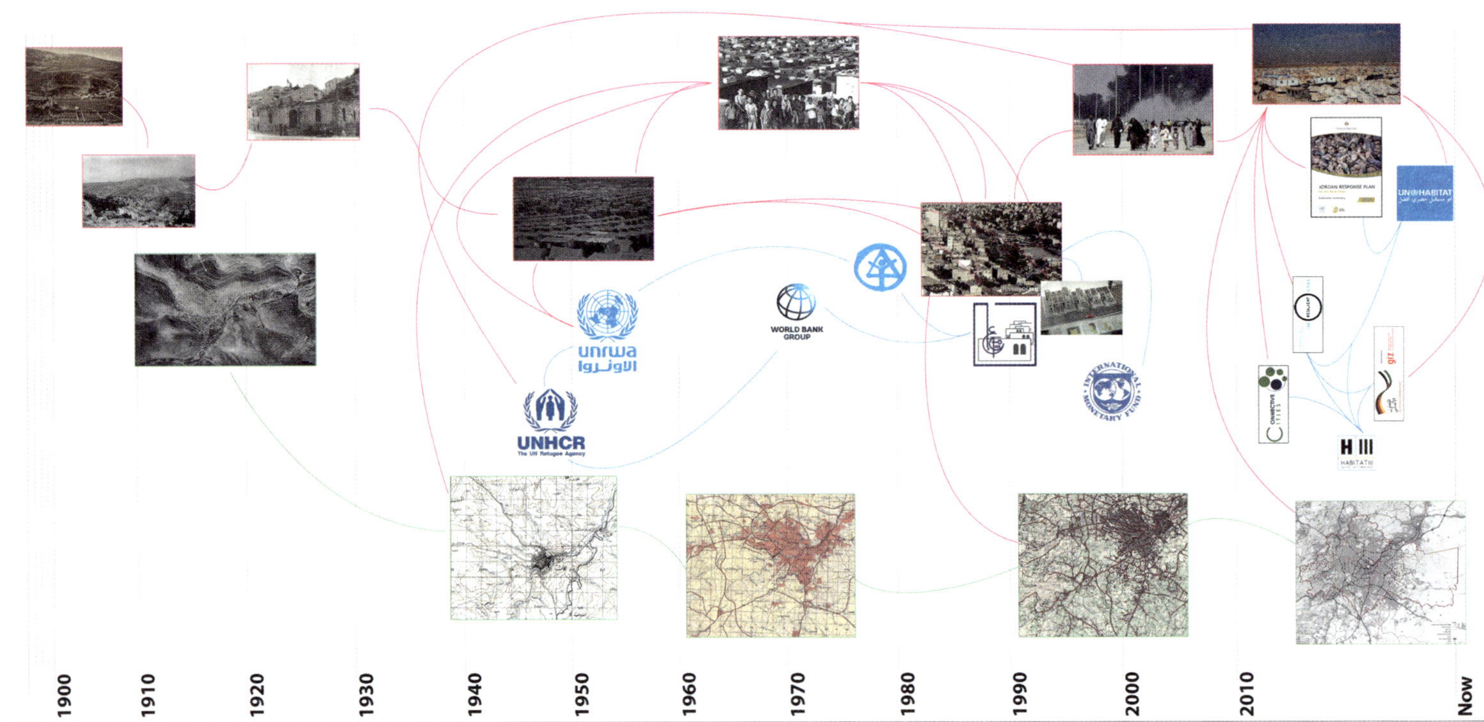

Figure 3 Urban growth of Amman (Qusay Amer, 2024)
Figure 4 The threefold interaction of factors influencing the urban fabric of Amman (Qusay Amer, 2024)

various groups involved. For example, the research in Amman indicates that refugees and asylum seekers also have different repertoires of resourcefulness and connections that they can mobilise or rely on (Simone 2004).

Mapping the Actor Network of a Green Infrastructure Project

In the illustrations (Figures 5 and 6) of the complex trans-scalar organisation and intergovernmental territorialisation involved in the rehabilitation of a public park in the eastern part of the city (Nadi Al-Sibaq Park), I aimed to show the interconnectedness of the many actors (local, national, and international ones). A hierarchical structure can be seen in the project's implementations, but this does mean that the whole process followed a pyramidal shape, as the decision-making, task assigning, and design processes were negotiated the entire time and between the different actors. If we consider the park to be the main critical zone and rely on the anamorphosis in projective geometry, as Farías and Paulos (2024) describe in reference to the work of Arènes, Latour, and Gaillardet (2018), we can visualise those dynamics as many smaller circles, each of which has its own critical junctures. Figuring out how to show these dynamics differently, as seen in the first sketches (Figure 7), I thought about time as an additional dimension and tried to address the simultaneity of the happenings—in this case, negotiations, conflicts, and production of space. I drew the size of the circles proportionately to the significance of each actor involved in the project.

As mentioned previously, to cope with the influx of refugees, Amman followed many strategies. One of these was to rely on the structure of existing international city networks that offered technical support and access to funds. It is worth mentioning that, in these new types of city networks and partnerships, 'the centrality of powerful private actors leads to suspicion that *hidden agendas* exist beyond their commitment' (Allegretti 2019, 44). For this research, the 100RC network was chosen for further examination. Amman was selected to join the 100RC network in 2014, which provided technical support for the development of the Amman Resilience Strategy 2017 and access to fund resources. This strategy encompasses diverse topics, such as energy, building, mobility, waste management, water, and urban planning. Climate change, migration, and refugees were integrated as cross-cutting themes across all subjects (Greater Amman Municipality 2017).

To address the topic of migration, the city initiated different workshops with various local, national, and international actors between 2014 and 2017. The main partner that helped organise this was the International Rescue Committee, which also collaborated with various local organisations, including Injaz and the Al-Osar society. They all worked to conduct a series of workshops incorporating the perspectives of refugees, migrants, and locals into this strategy. The second step after publishing the strategy was securing the funds through private connections. Parallel to the launching of the strategy, the German Development Cooperation (GIZ) started the project *Green Infrastructure in Amman – Improving Living Conditions of Disadvantaged Areas in Amman's Green Infrastructure*. As stated in the GIZ fact sheet about the project, the project contributes to Sustainable Development Goals (SDG) 11 and 13 and is locally integrated into GAM's wider strategy to develop walkability, access to public transport, and address the effects of climate change through the C40 and 100RC initiatives. Through the Jordan Response Plan, which defines a multi-year framework for steering and coordinating the aid response to Syrian refugees and vulnerable Jordanians (MoPIC 2015), both groups are guaranteed access to the cash-for-work programme. This is guided by the Jordan Response Platform for the Syria Crisis, under the leadership of the Government of Jordan.

The national bodies involved in the project have the role of setting the framework and general policies. However, we observed a direct link between international organisations and the municipality, which could be described as a scale jump, according to Davidson et al. (2019). Adding this to the autonomy

Nadi Al -Sibq Park - Timeline

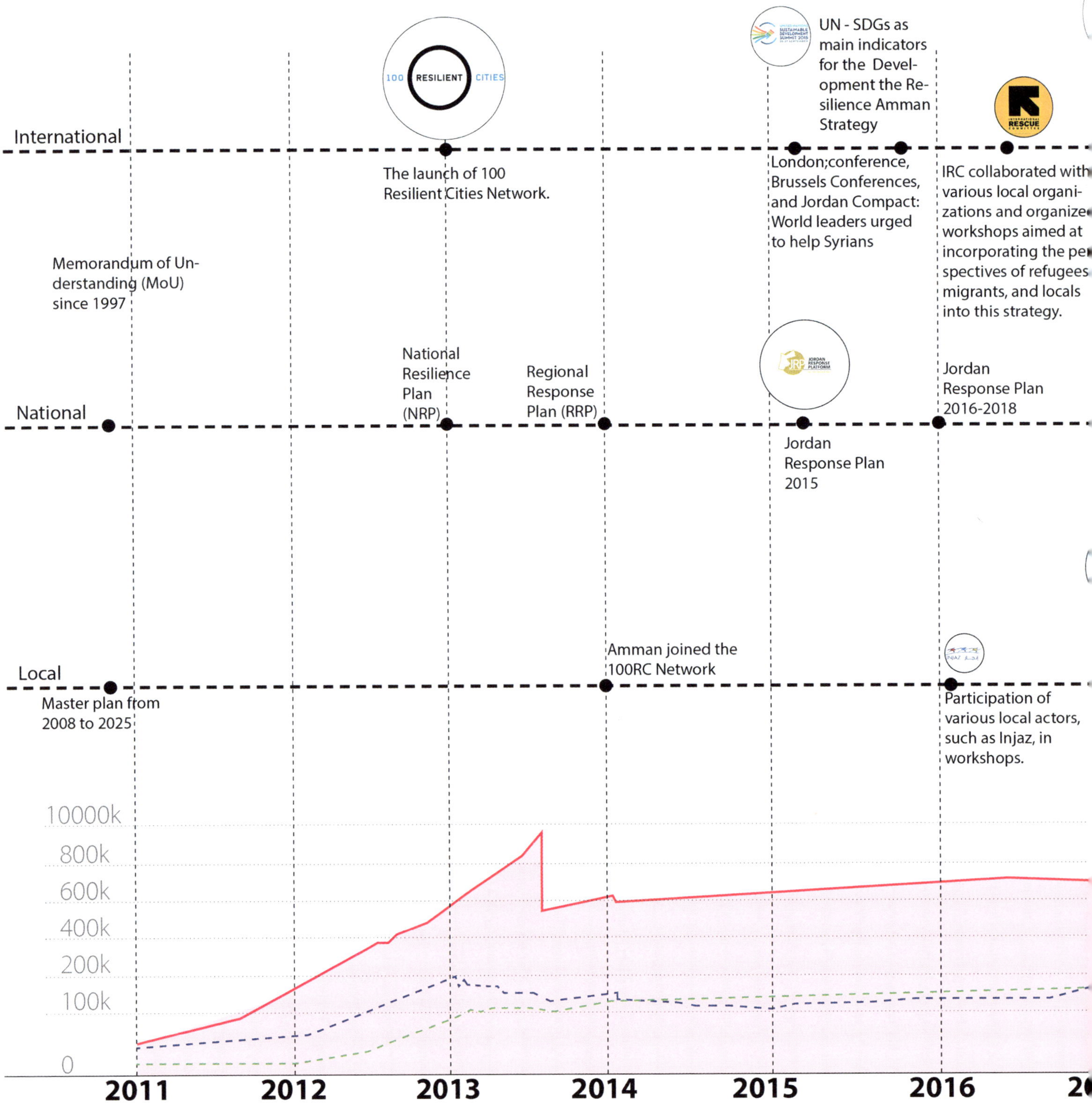

Figure 5 Nadi Al Sibaq Park timeline: initiatives, policy papers, donors, and experts (Qusay Amer, 2024)

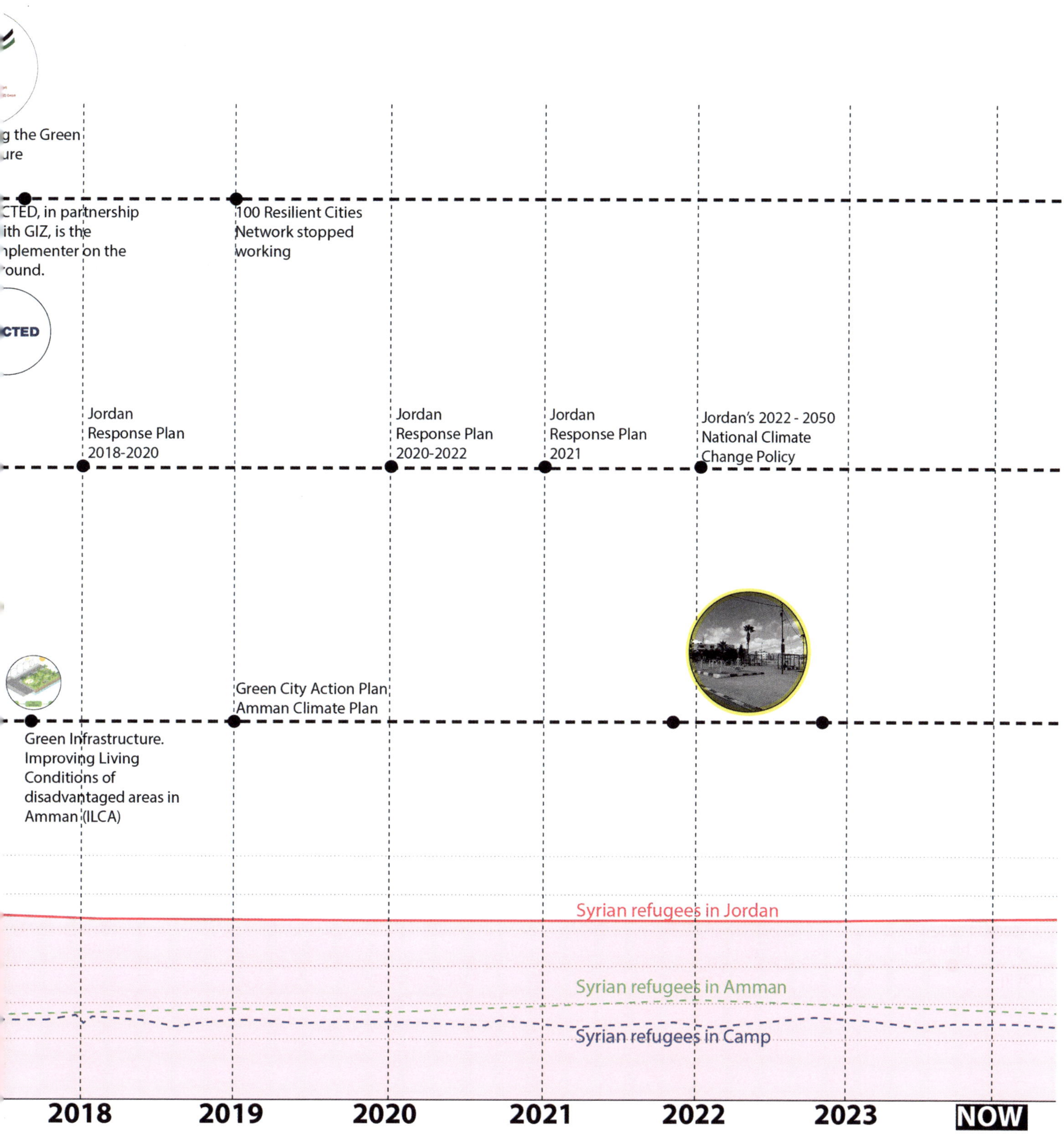
g the Green
ure
CTED, in partnership
ith GIZ, is the
nplementer on the
round.
CTED
100 Resilient Cities
Network stopped
working
Jordan
Response Plan
2018-2020
Jordan
Response Plan
2020-2022
Jordan
Response Plan
2021
Jordan's 2022 - 2050
National Climate
Change Policy
Green City Action Plan
Amman Climate Plan
Green Infrastructure.
Improving Living
Conditions of
disadvantaged areas in
Amman (ILCA)
Syrian refugees in Jordan
Syrian refugees in Amman
Syrian refugees in Camp
2018
2019
2020
2021
2022
2023
NOW

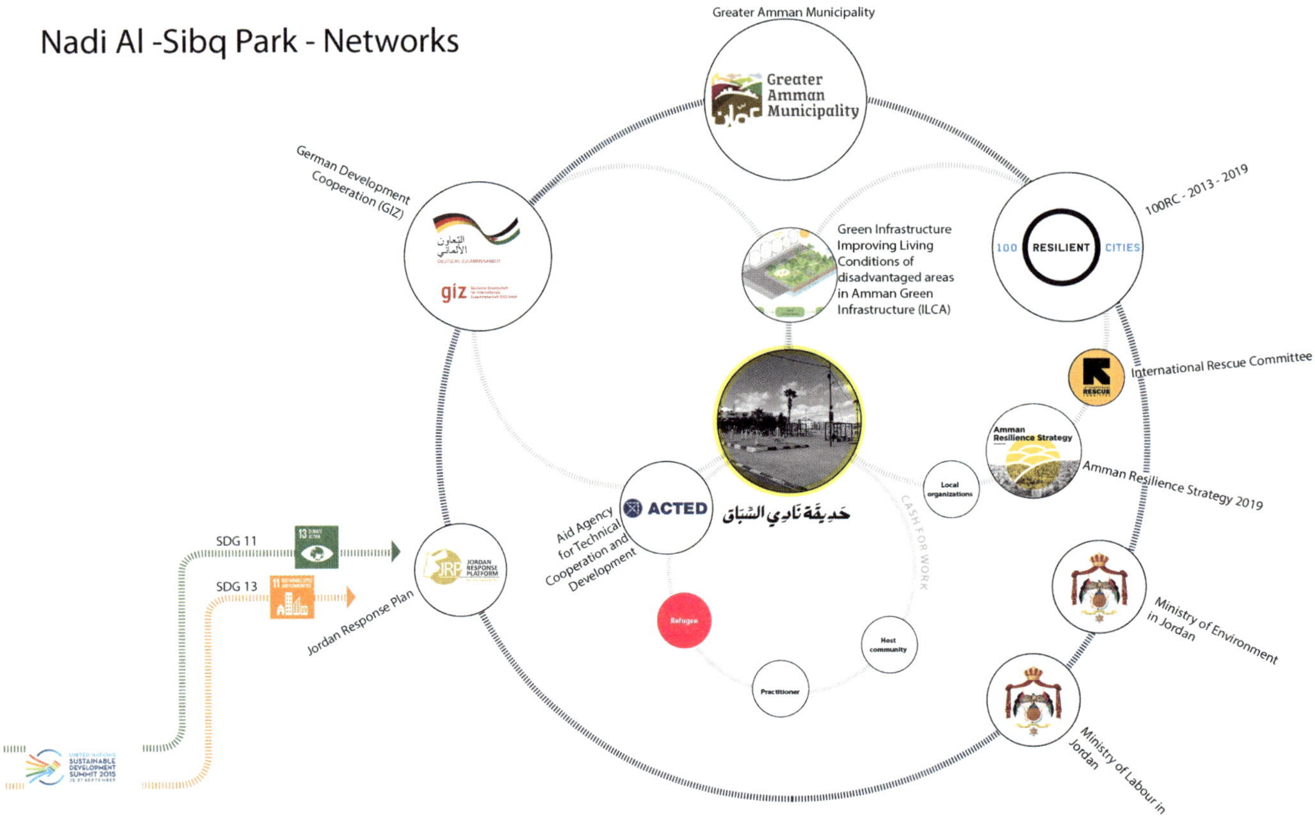

Figure 6 Nadi Al -Sibq Park: The complex trans-scalar organisation and intergovernmental territorialisation involved in the project (Qusay Amer, 2024)

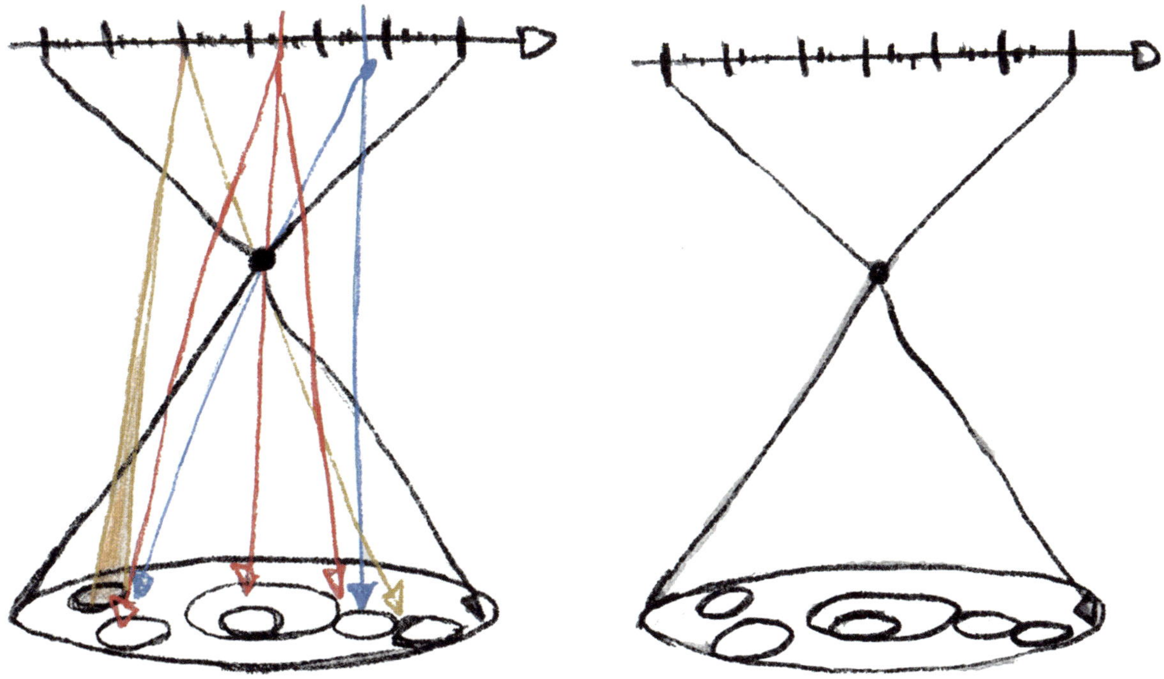

of GAM in terms of urban planning policies, city networks could be observed as catalysts for politics in most contemporary international institutions beyond states, and this highlights the potential of cities to shape international politics (Davidson et al. 2019).

The network mapping in both versions (the timeline and the trans-scalar) revealed how the project was implanted. Nevertheless, the change to the trans-scalar representation helped shed light on the negotiations that happened either simultaneously and affected the decision-making processes that impact the production of space; it also showed the role of each actor and their resources.

Final Reflections

Mapping helped address the challenge of movement always being spatial and temporal at the same time. The methodological approaches used to map time and space in this work attempted to include patterns beyond basic chronology. The timeline became mapping in the following circumstances: a) when the chapter connected historical and spatial events that happened in a different period of time, showing their interconnectedness; b) when, during the network mapping, we emphasised the different relationships between the actors involved in the production of space process and when, before, the strict linear dynamics did not fully exist and neither did the multilevel mapping of the actors; and c) when incorporating social, political, and economic factors into the mapping, which are not based on a temporal axis, was found to supplement or substitute time in numerous instances. This helped with a nuanced examination of space-time, adding non-linearity and covering dynamics, to understand the relational nature of the spatial processes.

Figure 7 The first sketches of the complex trans-scalar organisation and intergovernmental territorialisation involved in the project, with the timeline as an additional dimension (Qusay Amer, 2024)

In addition, I would like to highlight the features that I considered to transversally connect our work and approach to mapping as a methodology and an epistemological stance. *First*, directly related to the temporal horizons of spatial studies, I found that the historical scale matters as much as the geographical scale. The mapping of Amman's contemporary urban development's interlacing with rentier (neoliberal) economic manoeuvres demanded also looking at the migration movements of the last century, as their imprint on the city's fabric and identity politics are relevant to the present day. *Second*, the polyvalence of mapping as a heterogeneous process generated incommensurable images. The map of actors' trans-scalar agency in spatial interventions of Amman pertained to mapping of the plan of the park to be built. It identified, profiled, and put into relation the subjects, elements, and capital —as well as the spatial products and intervention stemming from them—that set and keep policies in motion. *Third*, as architects, sketching and visualising the received information during fieldwork is not just a procedural matter but an integral component of presenting complex realities in understandable forms. The development of the drawings was not based on aesthetics in the conventional sense, but rather on a need to present the vast amount and complexity of the data in a simple and effective way. *Fourth*, and lastly, Weidenhaus (2023) stresses the need to integrate space and time in sociology research. Based on that, I argue that studying the spatialities of refuge intrinsically requires space-time attention, as migration unfolds in both space and time (Vanyoro 2024). The easier way to research refugees' spatial practices —or governments' and aid organisations' asylum governance spatial interventions—would be to either fix time and look at the multiple refuge spaces or to fix space and look at the refuge spatialities through chronological lenses. Using neither variable fix, I hope to better attend to the complexities of migration and forced displacement as processes that never stop, neither spatially nor temporally.

Acknowledgements

The author would like to thank all of the anonymous refugees who participated in the interviews and to extend gratitude to the organisational team of the Data Justice for Refugees workshop. I also thank my teammate, Francesca Ceola, for all the discussions and ideas and everyone who was involved in developing these ideas.

Funding Statement

The research was funded by the German Research Foundation (Deutsche Forschungsgemeinschaft). Projektnummer 290045248 – SFB 1265.

Image Credits

Figure 1: Qusay Amer
Images (from left to right):

1 Amman, 1867–1914, Album 404 [1867–1914]; source: Album 404 [c 1867-c 1914], British Library, EAP644/1/37, https://eap.bl.uk/archive-file/EAP644-1-37 (accessed 12 August 2024).
2 The Roman Theatre area, Amman, 1920s; source: Circassian Voices; http://circasvoices.blogspot.com/2014/03/graphic-chronicle-of-establishment-of.html (accessed 12 August 2024).
3 Prime Minister's House, 1921, Amman; source: Department of the National Library; https://www.nl.gov.jo/Default/Ar (accessed 12 August 2024).
4 Palestinian Urban Refugee Camps, Amman, Jordan; source: UNRWA Audio Visual Archive.
5 Refugees in the camps soon after Nakba; source: Palestine Remember; https://www.palestineremembered.com/Articles/In-The-Refugee-Camps/Picture113521.html (accessed 12 August 2024).
6 Amman, 1992; source: Department of the National Library; http://www.jordanhistory.com/kharja/photos/collection1_en.php?id=7&page=7694 (accessed 12 August 2024).
7 Local residents flee Basra in southern Iraq, following the US-led invasion, on 28 March 2003. Agence France-Presse (AFP); https://www.middleeasteye.net/opinion/iraq-war-us-campuses-militarised-forgot-war-on-terror (accessed 12 August 2024);
8 Za'atari refugee camp, Jordan, in November 2012; source: UNHCR/Brian Sokol; https://reliefweb.int/report/jordan/jordans-zaatari-refugee-camp-10-facts-10-years-enar (accessed 12 August 2024).

Figure 2: Qusay Amer
Image: East Wahdat Upgrading Programme, Amman; source: Aga Khan Award for Architecture; https://the.akdn/en/how-we-work/our-agencies/aga-khan-trust-culture/akaa/east-wahdat-upgrading-programme (accessed 12 August 2024).

Figure 3: Qusay Amer
Images (from left to right):

1 A historical aerial photograph of the Amman Citadel and its surroundings in 1918; source: Archives of the Bavarian State, 1918; https://www.tandfonline.com/doi/full/10.1080/00758914.2021.2019412 (accessed 12 August 2024);
2 Amman 1–50,000, 1946–1947; source: Ifpo Library Amman, Royal Jordanian Geographic Centre.
3 Amman edition II, 1960–1961; source: Ifpo Library Amman, Jordanian Department of Land and Surveys;
4 Amman 1–100, 1993; source: Ifpo Library Amman, Royal Jordanian Geographic Centre;
5 Amman, 2022, Amman Spatial Profile; source: UN-Habitat.

Figure 4 , 5, 6, 7: Qusay Amer
Images: see above.

References

Ababsa, Myriam. 2012. 'Public Policies Toward Informal Settlements in Jordan (1965–2010)'. In *Popular Housing and Urban Land Tenure in the Middle East: Case Studies from Egypt, Syria, Jordan, Lebanon, and Turkey*, edited by Myriam Ababsa, Baudouin Dupret, and Eric Denis, pp. 259–282. Cairo: American University in Cairo Press.

Al Hamarneh, Ala. 2002. 'The Social and Political Effects of Transformation Processes in Palestinian Refugee Camps in the Amman Metropolitan Area (1989–99)'. In *Jordan in Transition*, edited by Georges Joffe, pp. 172–289. London: Hurst & Company.

Allegretti, Giovanni. 2019. 'Cities, Citizens and Demodiversity: An Overview of Two Generations of Cities Networks'. In *Rethinking the Ecosystem of International City Networks. Challenges and Opportunities*, edited by Agustí Fernández de Losada and Hannah Abdullah, pp. 39–48. Barcelona: Global Cities Programme, CIDOB.

Al Tal, Raed, and Hala Ghanem. 2019. 'Impact of the Syrian Crisis on the Socio-Spatial Transformation of Eastern Amman, Jordan'. *Frontiers of Architectural Research* 8(4): pp. 591–603.

Arènes, Alexandra, Bruno Latour, and Jérôme Gaillardet. 2018. 'Giving Depth to the Surface: An Exercise in the Gaia-Graphy of Critical Zones'. *The Anthropocene Review*, 5(1): pp. 120-135.

Brun, Cathrine, and Anita Fábos. 2015. 'Making Homes in Limbo? A Conceptual Framework'. *Refuge* 31(1): pp. 5–17.

Chatty, Dawn. 2010. 'Circassian, Chechnyan, and Other Muslim Communities Expelled from the Caucasus and the Balkans'. In *Displacement and Dispossession in the Modern Middle East*, edited by Dawn Chatty, pp. 91–133. The Contemporary Middle East. Cambridge: Cambridge University Press

Daher, F. Rami. 2011. 'Prelude. Understanding Cultural Change and Urban Transformations: Qualifying Amman: The City of Many Hats'. In *Villes, Pratiques Urbaines et Construction Nationale en Jordanie*, edited by Myriam Ababsa and Rami Farouk Daher, pp. 65–89. Beyrouth: Presses de l'Ifpo. https://doi.org/10.4000/books.ifpo.1725.

Davidson, Kathryn, Lars Coenen, Michele Acuto, and Brendan Gleeson. 2019. 'Reconfiguring Urban Governance in an Age of Rising City Networks: A Research Agenda'. *Urban Studies* 56(16): pp. 3540–3555. https://doi.org/10.1177/0042098018816010.

Department of Statistics. 2015. 'Jordan in Figures—National Census'. Amman, Jordan: Department of Statistics.

Farías, Ignacio, and Julio Paulos. 2024. 'Actor-Network Theory as a Theory of Space'. In *Handbook of Qualitative and Visual Methods in Spatial Research*, edited by Anna Juliane Heinrich, Séverine Marguin, Angela Million, and Jörg Stollmann, pp. 33–44. Bielefeld: transcript Verlag. https://doi.org/10.14361/9783839467343-003.

Greater Amman Municipality. 2017. *Amman Resilience Strategy. Resilient Cities Network*. Accessed 12 August 2024: https://resilientcitiesnetwork.org/downloadable_resources/Network/Amman-Resilience-Strategy-English.pdf.

Ghanem, Hala. 2022. *Public Outdoor Space and Refugees' Socio-Cultural Integration: Understanding Refugees' Agency within a Network of Actors in Amman, Jordan*. PhD dissertation, School of Landscape Architecture, University of Sheffield.

Hamed-Troyansky, Vladimir. 2017. 'Circassian Refugees and the Making of Amman, 1878–1914'. *International Journal of Middle East Studies* 49(4): pp. 605–623. doi:10.1017/S0020743817000617.

Heinrich, Anna Juliane, Séverine Marguin, Angela Million, and Jörg Stollmann. 2024. 'Introduction. Spatial Research from an Interdisciplinary Perspective'. In *Handbook of Qualitative and Visual Methods in Spatial Research*, edited by Anna Juliane Heinrich, Séverine Marguin, Angela Million, and Jörg Stollmann, pp. 9–16. Bielefeld: transcript Verlag. https://doi.org/10.14361/9783839467343-001.

Löw, Martina. 2016. 'Preface'. In *The Sociology of Space*, English ed., pp. vii–xxi. New York: Palgrave Macmillan.

Massey, Doreen B. 2005. 'Opening Propositions'. In *For Space*, pp. 9–15. London, UK: Sage Publications.

Ministry of Planning and International Cooperation. 2015. *Jordan Response Plan for the Syria Crises.*

Munif, Abd ar-Rahman. 2018. سيرة مدينة: عمان في الأربعينيات. Beirut: المؤسسة العربية للدراسات و النشر & دار التنوير للطباعة و النشر.

Ng-Chan, Taien. 2015. 'Mapping Out Patience: Cartography, Cinema and W.G. Sebald'. *Humanities* 4(4): pp. 554–568. https://doi.org/10.3390/h4040554.

Pelger, Dagmar. 2022. 'Kleine Geschichte kooperativer Planungsweisen'. In *For Dis-Closing Separate Space. Kartierungen Kooperativer Planungsarbeit Am Cotti, Am Kotti Und Am Hermann- Platz In Berlin, 4,* edited by Dagmar Pelger, pp. 1–5. Berlin: Universität der Künste.

Pelger, Dagmar, Emily Kelling, and Jörg Stollmann. 2021. 'Multiscalar Mapping'. In *Handbook of Qualitative and Visual Methods in Spatial Research*, edited by Anna Juliane Heinrich, Séverine Marguin, Angela Million, and Jörg Stollmann, pp. 247–262. Bielefeld: transcript Verlag.

Pisarevskaya, Asya, and Peter Scholten. 2022. 'Cities of Migration'. In *Introduction to Migration Studies*, edited by Peter Scholten, pp. 249–262. IMISCOE Research Series. Cham: Springer. https://doi.org/10.1007/978-3-030-92377-8_16.

Shami, Seteney. 2007. 'Amman is Not a City: Middle Eastern Cities in Question'. In *Urban Imaginaries: Locating the Modern City*, edited by Aley Cinar and Thomas Bender, pp. 208–235. Minneapolis: University of Minnesota Press.

Simone, AbdouMaliq. 2004. 'People as Infrastructure: Intersecting Fragments in Johannesburg'. *Public Culture* 16(3): pp. 407–429.
Tabačková, Zuzana. 2021. 'Site Visits'. In *Handbook of Qualitative and Visual Methods in Spatial Research*, edited by Anna Juliane Heinrich, Séverine Marguin, Angela Million, and Jörg Stollmann, pp. 199–212. Bielefeld: transcript Verlag.
United Nations Human Settlements Programme. 2022. 'Amman Spatial Profile'. *City Profile. Urban Planning & Infrastructure in Migration Contexts*. Amman: UN-Habitat.
Vanyoro, Kudakwashe. 2024. 'Chronopolitics: Decolonising African Migration Studies, Critical African Studies'. *Critical African Studies* 16(3): pp. 399–414. doi:10.1080/21681392.2024.2387554.
Weidenhaus, Gunter. 2023. 'Social Space-Time: On the Concept of Social Space-Time and Its Empirical Relevance to Biography'. *Civic Sociology* 4(1): article 57492. doi: https://doi.org/10.1525/cs.2023.57492.

Immersive Mapping

Jamie-Scott Baxter,
Lus Constantin,
Séverine Marguin

Mistaking the Map for the Territory in the Botaniverse

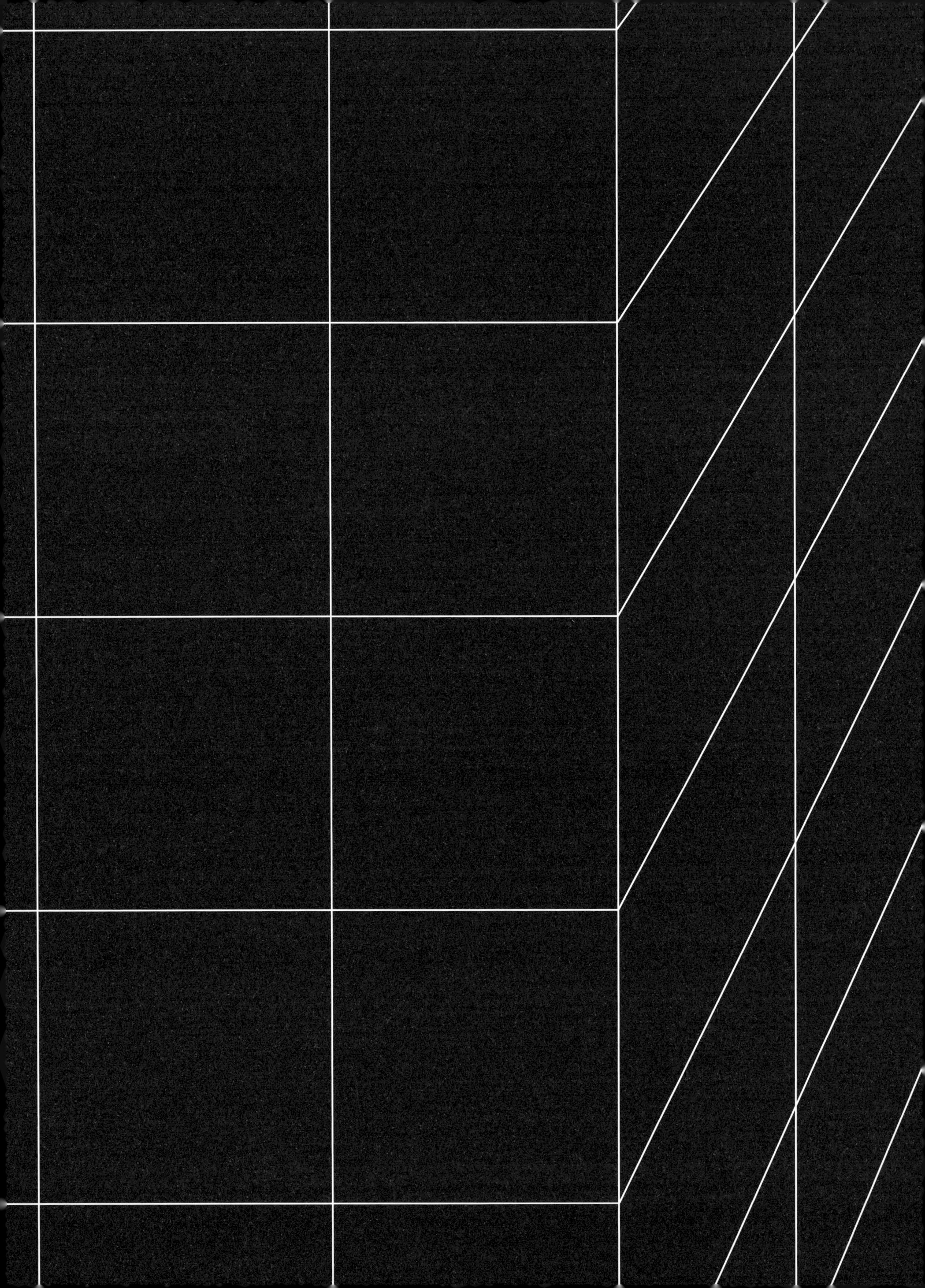

This chapter reflects upon a mapping experiment at the intersection of digital arts, architecture, and social sciences. It considers the potential of virtual reality technology as an immersive and multimodal mapping tool for time-sensitive spatial research and analysis. In the ambiguous realm of the so-called metaverse, we designed the Botaniverse[1] as an immersive digital environment equipped to display diverse multimodal media. It tells the story of the production of knowledge and space at botanical gardens. The experiment was based on empirical data collected during a three-year research project on the spatial refiguration of botanical gardens (Baxter and Marguin 2024).

1 Accessible here: www.monaverse.com/spaces/refigbo.

Our main purpose was to experiment with ways of capturing time in spatial mappings, which otherwise tend to be relatively static, or at least reproduce a singular temporality. Arguably, the atemporality of maps contributes to their performance of hegemonic power (Baxter et al., editorial, this volume). Building on the hybrid mapping methodology (Baxter et al. 2024) that we have developed over the past years and used in the current research project, we took on the challenge of considering multiple spatialities (Löw and Knoblauch 2022) alongside multiple temporalities (e.g., Chakrabarty 2018) that constitute more-than-human social worlds, exploring how to integrate time and space into our method. This meant complexifying the dominant narratives of human history, which is often viewed as sequential, singular, and progressive, and instead meticulously attending to the ways in which different times and spaces are entangled and created with one another across scales. To this end, we expanded the practices of spatial mapping to integrate multiple modalities, especially time-based media (e.g., video, animation, sound, etc.). Using a digital three-dimensional (3D) environment allowed for an immersive experience of the research material, thereby enhancing the 'spatiality' in the spacetime mapping.

In this paper, *first*, we discuss why integrating many spaces with many times in mapping protocols is relevant. *Second,* we describe our design process for the Botaniverse, which is understood as an example of immersive mapping. *Third,* we show how immersive mapping allowed us to analytically grasp differentiated empirical spacetimes unfolding across and structuring botanical gardens in our cases.

Hybrid Mapping: Many Spaces, but Only One Time

The art installation Botaniverse was born out of a two-year funded research project investigating the ways in which modern institutions are socially and spatially structured and how historically contingent spatial structures shape and restrict processes of change. The project developed from the premise that, while social change is inherently spatial, it might not follow time's arrow as concepts like transformation tend to imply. Instead, it is a non-linear, conflictual process taut with power dynamics. Theoretically, this challenges modern concepts of transformation and progress; more broadly, it troubles the unilinear path of ever-expanding globalisation. It is understood here as the *refiguration of spaces* (Knoblauch and Löw 2024; Löw and Knoblauch 2022).

Botanical gardens are paradigmatic modern institutions that present a case well-suited for studying the refiguration of spaces. On the one hand, they are engaged in scientific knowledge production, including the extraction and exchange of plants as technoscientific objects, while on the other, they are charged with the mission to 'educate' an urban and historically European public on the 'exotic' botanical wonders of the world—typically taken from one part (not Europe) and relocated to another (Europe). This familiar urban institution, often simultaneously a museum and public space, has historically shaped nature-culture relations and served to define what constitutes legitimate knowledge on 'Nature'—a central consideration of modernity (e.g., Barad 2007; Descola 2005; Latour 1993). Our research on Berlin's Botanical Garden and the Royal Botanical Garden Edinburgh showed that, at the beginning of the twenty-first century, botanical gardens have been repositioning themselves in relation to what can be understood as a changing conservation

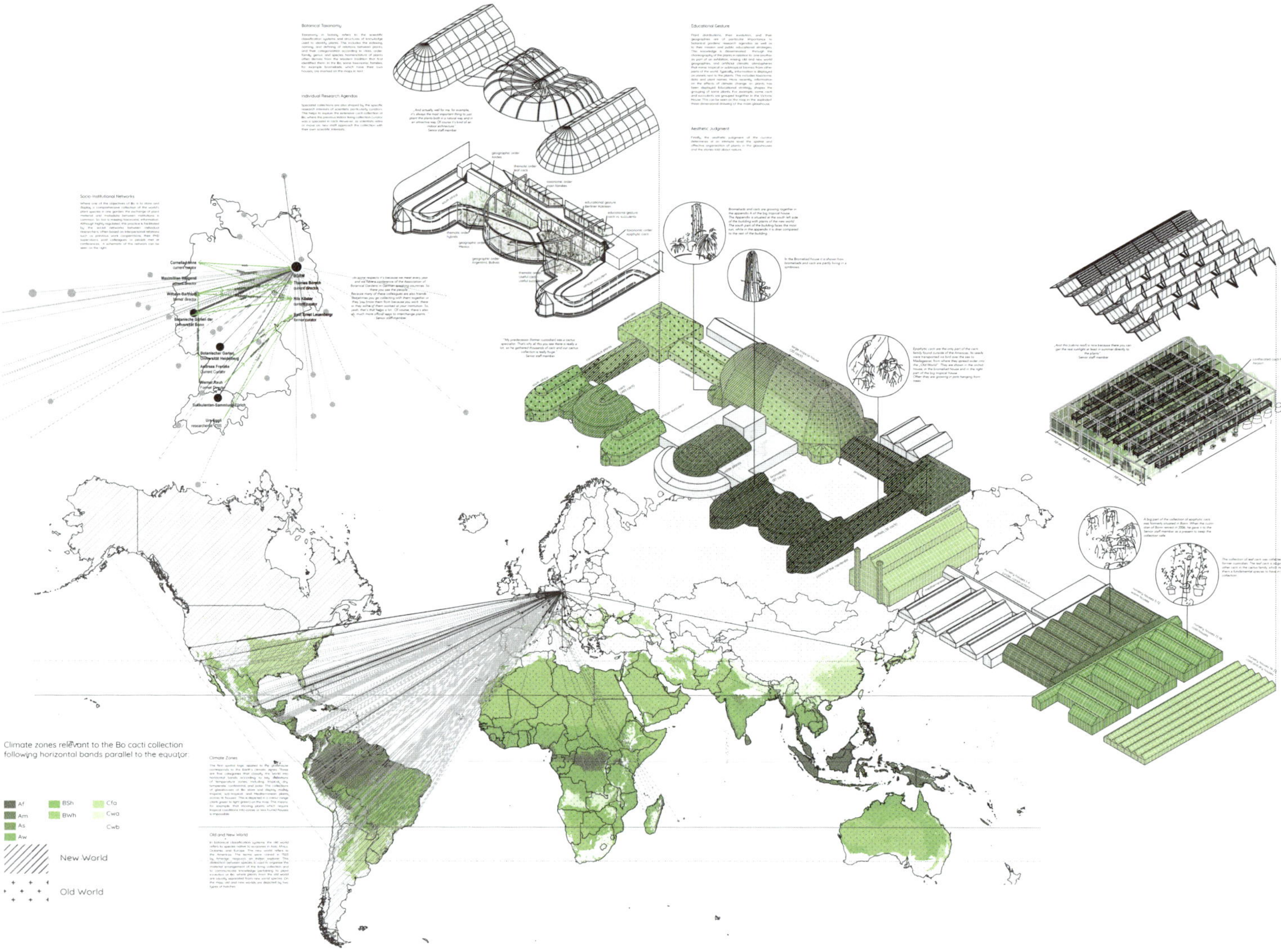

Figure 1 Hybrid mapping of the glasshouse in the Botanical Gardens in Berlin (Julius Frittkau and Zoe Hochstein-Morran, 2021)

regime (Adams 2013; Braverman 2023), which is itself shaped by contemporary material-discursive forces of climate change and decolonisation. Our analysis revealed that the historic spatial structure associated with the ways in which knowledge in gardens and glasshouses is spatially ordered constrains the possibility of social change in relation to these forces—a process of spatial refiguration rather than sequential transformation. In the context of modern institutions that deal explicitly with nature-culture relations, we may understand these spatial orders as specific 'stagings of nature' caught between conflicting logics of fixity and change (Baxter and Marguin 2024). Taking this a step further, the research revealed how botanical gardens are not only shaped by a conservation regime but are also powerful actors within it that disseminate highly networked spatial knowledge of nature, knowledge underpinned by Western epistemology, and modern technoscientific practice. This led to the understanding that botanical gardens are seemingly mobile spaces that are themselves in circulation over the world, imitating nature and their doppelgängers in many ways. We refer to this spatial logic as 'mimetic' to account for the spread of spaces that are simultaneously the same *and* different. In this case, mimetic spaces reproduce specific Western constructions of nature and conservation that shape the material-discursive boundaries that determine human and nonhuman life (Baxter and Marguin 2025).

The research consisted of ethnographic site visits, which included photographs, video, and sketches, and expert interviews with main actors (director, custodians, gardeners, and scientists). Parallel to the research, we led two courses with master's students at the Technische Universität Berlin, during which the initial mappings presented below were made (Figure 1 for an example). We applied the hybrid mapping methodology (Baxter et al. 2024) to render visible and analyse the translocal spatial assemblages, spatial structures, diverse actors, and material–discursive infrastructures at work in the staging of nature and knowledge at the two botanical gardens. Hybrid mapping is an interpretative method of analysis based on the visual synthesising of heterogeneous data (quantitative/qualitative; multimodal) that combines interdisciplinary ways of knowing from design practice and social science research. Initial operations include the spatial coding of qualitative interviews and ethnographic data into multiple drawings that relationally integrate spatial knowledge, spatial practices, and spatial structures into a joint spatial display (Marguin et al. 2024). These are followed by an iterative and abductive composition process that brings together single drawings into synthetic mappings, with the aim of drawing out the multiple spatial arrangements at work in these settings. Conceptual analysis is carried out both by visually working out points of congruence, distance, and conflict and by developing a legend and annotations in the mapping.

Hybrid mapping proved extremely fruitful for the data collection and spatial analysis and aided our understanding of progression in time through, for example, the incorporation of a sequence of spatial morphology drawings (Rekacewicz this volume). However, the method was less capable of drawing out the complex temporalities at work in the case, as the refiguration of space theory aims to consider. It relays much more on a singular and linear understanding of time. For this reason, we posed the following question considered in this article: *how might we better include multiple temporalities in the hybrid mapping method to grasp complex spatiotemporal processes, for example, by expanding our understanding of multimodality in mapping to include digital time-based media, such as video, animation, or sound*?

Spacetimes and Planetary Matters

As indicated above, the concept of staging nature deconstructs and describes the specific spatial arrangements of botanical glasshouses and gardens as a multiplicity of social and ecological forces undergirded by colonial histories and Western technoscientific practices. Staging nature implies both the

Siteplan Botanical Garden Berlin

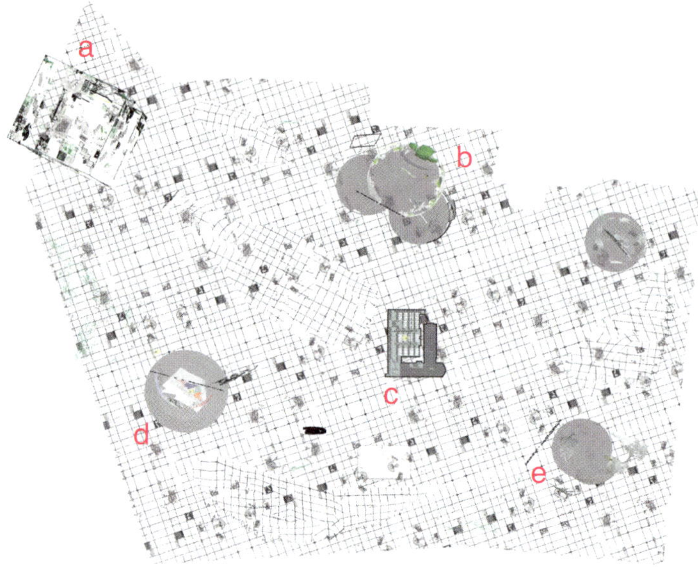

Ground Level Botaniverse

practices and knowledge that go into the spatial organisation of botanical gardens. It is a specific type of spatial arrangement acknowledging, on the one hand, the social construction of nature, carefully choreographed as 'wild' or 'pristine' and entangled with curatorial and scientific practices, and on the other hand, the performativity of nature as agential, lively, material nonhuman life.[2]

Correspondingly, and grounded in a relational understanding of space and time, this position temporarily suspends disbelief over the possibility that spacetimes are at once materially *real* and socially constructed (Baxter 2023). Space, in its multiplicity (Knoblauch 2022) and multiscalarity, is continuously performed in practice—practices that include nonhuman material-discursive planetary forces (Baxter 2023). The art installation Botaniverse aimed to experiment with ways to complexify this critical materialist spatial approach by confronting the relation of space and time in map making. As Massey explains, 'spatializing time (thinking in terms of spacetime) is thus also to propose a genuine multiplicity of trajectories' (Massey 2001, 259). This requires thinking of space and time as co-constitutive of each other (Massey 2005), thus raising questions regarding the extent to which 'specific spatial conditions are directly related to specific temporal conditions' (Weidenhaus 2017, 41, our translation). Thinking of the construction of space through and with the construction of multiple temporalities requires overcoming linear or even cyclical understandings of time (Chakrabarty 2000; Koselleck 2004, 2018). This goes hand in hand with the idea of the difference and simultaneity of times: 'For time, a concept of chronology with reference to the external world, one of historicity with reference to the social world and one of the experience of duration with reference to the inner-psychic world can be differentiated' (Weidenhaus 2017, 42, our translation). These differentiations of time have been critically tackled by postcolonial and feminist literature to contest the singular story of Western modernity (Jameson 2003), which has for too long served as a model for conceiving and analysing other societies (Mudimbe 1988). Attempting to think against the hegemony of time's linearity (progress) and singularity (history), which carries an implicit assumption of a Western progress orientation, requires meticulous and careful focus on complex entanglements

2 For an extensive discussion on the agency of nature and nonhumans and the performativity of reality, see, e.g., Coole and Frost (2010); Barad (2007); Haraway (1991); Bennett (2010); Latour (1993); Tsing et al. (2019); and for contrast, see Ernstson and Swyngedouw (2019); Gandy (2024).

Figure 2 Floorplan of the 3D space created for the Botaniverse (Jamie-Scott Baxter, Lus Constantin, and Séverine Marguin with the assistance of Mara Oppermann, 2023)

between different temporal and spatial scales—or, as Chakrabarty (2000, 109) puts it, *drawing* attention to 'a plurality of times existing together, a disjuncture of the present with itself'.

The notion of a disjunctured present is particularly pertinent when it comes to making sense of the contemporary socioecological crisis we face. Chakrabarty (2018) emphasised that, to grasp the specificity of our current conditions—or rather what is now required of us—is to hold together at once at least two temporalities: the modern historical time of humanity and a much deeper nonhuman time of the planetary. The peculiar paradox of this situation is that the planetary was only able to be discovered through the history of modernity and globalisation. We draw on this idea of simultaneously overlapping conflicting temporalities of deep time, historical time, and lifetime as 'complex coevalness' (Fabian 1983).

The Botaniverse

To include multiple different and overlapping temporalities and grasp complex spatiotemporal figurations, it became clear that we needed to extend the hybrid mapping method by integrating time-based media into the mapping process. This led to extending our interdisciplinary collaboration between sociology and architecture to digital art and architecture in order to integrate the experience into digital media, including 3D modelling and rendering, film, sound, and animation. We soon decided that virtual reality technology may be suitable, not only to include time-based media in the expansion of our hybrid mapping protocol, but also to create a new spatial experience reminiscent of the ways in which botanical gardens are themselves mimetic and illusory spaces that are woven together from fragmented miniaturised landscapes abstracted from place and time. We aimed to recall the mimetic logics inherent in botanical gardens themselves (see Baxter and Marguin 2025). Providing maps—as miniaturised landscapes that could be entered into—allowed us to play on the idea of purposefully 'mistaking the map for the territory' (Korzybski 1933), distorting the relation between reality and representation.

These criteria led us to the generic virtual reality technology of the 'metaverse'; specifically, the platform *Monaverse* was selected. This technology and platform was deemed suitable, as it is a dedicated 3D virtual environment used in many cases to display and trade multimedia digital artwork, including non-fungible tokens, videos, and animations. It is also free of charge, thus helping reduce barriers to access. Immersive virtual reality environments can be created using CAD modelling and rendering software and can be easily uploaded to the *Monaverse* platform. Importantly, once uploaded, spaces can host multiple and diverse time-based media, which is thereby accessible to anyone with a web browser, stable internet connection, and rudimentary point-of-view gaming experience. The virtual environment allows for collaboration and interaction with other people via avatars. In the following sections, we describe the steps taken to construct a virtual 3D environment that can integrate the time-based multimodal media collected and created during the research. We refer to this new digital space as the Botaniverse.

Step 1. Disassembling (Hybrid Mappings)

In the first step, we dismantled the hybrid mappings made during our research to isolate the heterogeneous elements from which they were made. This allowed us to reflect upon and recode the relations between elements using *mises-en-correspondences* (Figure 2), thereby allowing for further in-depth analysis. During this process, four stories emerged that would go on to serve as a central spatial organisational structure for the design and curation of the Botaniverse. These stories include 'Landscapes Locked', which shows how conservation and heritage protection stratify landscape spaces, thus locking them in time and thereby structuring how change can occur; 'Taxonomy Unpicked', which describes the powerful ways in which botanical knowledge is deeply inscribed into space through taxonomy; 'Archives Distributed,' which

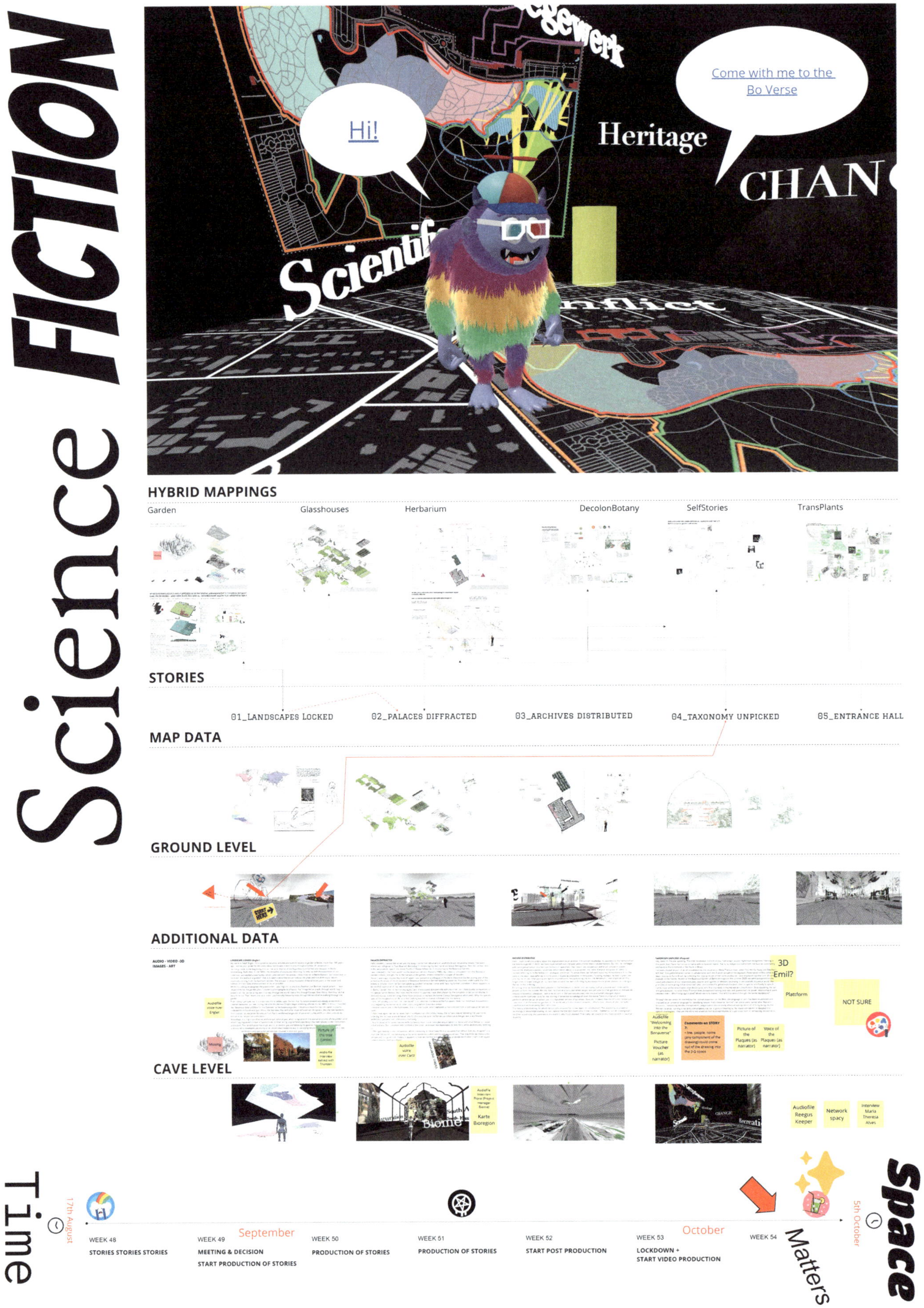

Figure 3 *Mises-en-correspondence* between elements of the initial mappings (Jamie-Scott Baxter, Lus Constantin, and Séverine Marguin with the assistance of Mara Oppermann, 2023)

Figure 5 Photographic view within the Botaniverse (Jamie-Scott Baxter, Lus Constantin, and Séverine Marguin with the assistance of Mara Oppermann, 2025)

BGBM
Heidelberg
Andreas Franzke
Werner Rauh
ZSSv
F. Ritter

visually presents how herbaria are seemingly de- and re-territorialised through processes of digitalisation; and 'Palaces Diffracted,' which describes how spaces of botanical glasshouse in the Global North are mimetically constructed in relation to spaces of nature in the Global South. These spatial stories were restaged in the Botaniverse through the integration of mappings, photos, videos, animations, and voiceovers and through the inclusion of raw data, such as interview recordings and go-along video excerpts.

'Landscapes Locked'

'Landscapes Locked' is centred around the social figure of Adolf Engler, the founder of the Berlin Botanical Garden more than 100 years ago. Together with architect Alfred Koerner, their vision was to recreate in Berlin 'the world in a garden', displaying 'exotic' plants collected from across the world. The botanical garden layout was based on Engler's world travels and the botanical knowledge of the time, both facilitated by colonialism. During this era, the German Empire began annexing colonies in Africa. Engler founded the Botanical Central Office for the German Colonies at the Botanical Garden. It became a nexus for the circulation of plants collected from South and East Africa (e.g., Namibia). Many specimens collected from other colonies were exchanged to enhance the respective collections of botanical gardens in other imperial European nations, for example, Kew Gardens in London and the Jardin Botanique in Paris (Baber 2016; Kaiser 2022).

'Taxonomy Unpicked'

'Taxonomy Unpicked' shows a resistance to hegemonic naming practices by reproducing the 2020 art installation *You Will Go Away One Day But I Will Not* by visual artist Maria Thereza Alves in the Berlin Botanical Garden. For this installation, Alves gathered plant names in collaboration with the Guaraní people of the Jaguapiru Reservation in Dourados in Mato Grosso do Sul in Brazil and collected singing rituals for name attribution.

'Palaces Diffracted'

In 'Palaces Diffracted', the story is built around the more-than-human social figure of a cactus, who narrates the story of its own circulation and (dis)placement into the garden. Originally from Umuarama in southern Brazil, it was collected in the early nineteen-eighties. The cactus's collector, a botanist from the Botanical Garden in Bonn, brought it to the garden's nursery, where it stayed for a couple of decades. About five years ago, its journey began again. Circulating via the social-institutional networks of German-speaking botanical academia, the cactus landed in a specific glasshouse of the Berlin Botanical Garden following strict climatic, taxonomic spatial orderings as well as the distinction between old and new worlds. In our investigation of the plant arrangement within the garden, we were interested in deciphering the space-time implications of this botanical imagination of the world. In the botanical classification system, the old world refers to ecozones in Asia, Africa, Oceania, and Europe, whereas the new world refers to those in the Americas.

'Archives Distributed'

'Archives Distributed' is built around the story of herbaria and the technoscientific objects they host—that is, botanical vouchers. Vouchers are herbarium specimens that are pressed and dried samples of a plant. Systematic information is arranged around the dried plant extract; this information includes the name, the year and place of collection, a reference number in the herbarium catalogue, and so on. Vouchers are the product of serial translations from plant to data. They are conserved within the archives of the herbarium, simultaneously following and performing the taxonomic classification of botanical knowledge.

Figure 6, 7 Photographic views within the Botaniverse (Jamie-Scott Baxter, Lus Constantin, and Séverine Marguin with the assistance of Mara Oppermann, 2025)

Step 2. Reassembling (the Botaniverse)

The next step was to assemble the Botaniverse by playfully imitating techniques similar to those of real-life botanical garden spaces (e.g., mimesis, affective illusion, miniaturisation, and domination), which we learned through our research are full of colonial power relations (Baxter and Marguin 2025). For example, the ground plan for the new digital garden is based on the same geometry as that of Berlin's Botanical Garden (Figure 3). The surface texture is composed of ordnance gridlines and tree and plant symbols taken from (and deliberately referencing) the original maps used in the research phase. The routes and sequencing between each of the four stories were carefully choreographed. In this way, not only were miniature stories staged *in* the digital garden, but together, they *constituted* a fragmented digital replica of a botanical garden space featuring a polycontextural landscape of nature stories collected from afar and woven together in a new digital botanical garden: the Botaniverse.

Lying beneath this digital landscape is a subterranean world of virtual caves that house and display raw data collected during the research. Similar to including field notes or interview material in an essay, a network of dislocated subterranean spaces provides a deepening of the interpretative stories 'above ground' by giving the visitors access to some of the original data collected during the research, such as interview audio or go-along video material. Taking a cue from science fiction, the caves are instantly accessed via teleportals concealed in each of the stories above. Similar to point-of-view gaming, the portals are only revealed to visitors as their avatars delve deeper into the 3D environments. From the caves, another portal projects visitors back to the 'main entrance'. This creates a secondary circulation route, offering the possibility of alternative spatial and cognitive relations between stories—not dissimilar to how mapping renders spatial relations visible. The linear transmission of ideas that an essay would allow becomes distributed and ambiguous, provoking visitors (which might include collaborating analysts) to form new associations between multimodal data and, ultimately, new interpretations. The following section lays out new interpretations that the Botaniverse brought forth for our research on the refiguration of spaces in botanical gardens.

Botanical Gardens: Multiple and Overlapping Spacetime Figurations

Allowing for new associations and interpretations to be made jointly, the Botaniverse transcends its status as solely a creative space for science communication (akin to, e.g., an exhibition) and simultaneously becomes an analytical method. This has parallels to the 'hybridity' in hybrid mapping, which aims to 1) overcome and complexify the linearity in typical research processes by circling back data collection, analysis, and visualisation; 2) provide a means for intersubjective and collective analysis of data around maps; 3) hold heterogeneous data together in a 'joint spatial display' (Marguin et al. 2024); and 4) combine designerly ways of knowing with social scientific methods for new interpretations (Baxter et al. 2024). However, the use of a 3D environment combined with basic virtual reality in the Botaniverse allowed us to go further. It provided a digital space and platform for integrating multi-modal media into the 'joint spatial display', in this case, time-based media and not only different types of data but also data in different states of interpretation (e.g., raw audio from interviews with synthetic diagrams). On the one hand, the gaming interaction of virtual reality allowed for an immersive experience of the research material, thereby enhancing the 'spatiality' in the 'joint spatial display'. On the other hand, the ability to interact collectively in the 3D environment allowed for the collaborative interpretation of material over distance. In this way, the Botaniverse, simultaneously understood as a method and a space of display and communication, is like stepping into/through a map, further complexifying the map-territory relation. What might we call this type of mapping that does not leave any ambiguity to the spaces it creates (Willis and Fry this volume) and operates in? Such mappings embrace multiple modalities while holding in tension multiple types of data—a method that aims to display and trace the entanglement of multiple spaces and times and how they co-constitute one another. We suggest calling it *immersive mapping* (Figure 3).

Not only did immersive mapping shine a brighter light on the multiple spatialities overlapping and co-constituting botanical gardens; it also helped grasp the ways in which multiple temporalities are entangled in these spaces. Below, we illustrate the multiple spacetimes which offer new interpretation to our botanical garden research, they are: colonial spacetime, planetary spacetime, and perpetual spacetime (Figure 4). And with each, we highlight a critical juncture that ruptures spacetimes' smooth progression.

Colonial Spacetime

'Landscapes Locked' (see above), staged as two animated maps, shows the parallel between Engler's travels and the layout of the botanical gardens, illustrating how remnants of the garden's colonial past are continually projected into the present through the stratified spatial ordering and spatialised botanical knowledge (e.g., plant geography) inscribed in and constitutive of the garden layout. Positioning a further map in the ensemble brings to light an additional temporal dimension visualising how the garden's landscape design has been placed under Monument Protection (dating back to the mid-nineteen-nineties), and it depicts how spacetimes are fixed through the conservation regime of heritage protection. In this spatial fixing of time, visitors follow Engler's gaze over 'Nature'—that is, the gaze of a Western European scientist who further developed taxonomy as a universalist modernist way of classifying plants. This is nature staged from a particular perspective and reproducing a spacetime in which European imperialism, colonialism, and early botanical science are deeply woven together. This story depicts how a spatial figuration strongly associated with territorial expansion and the Western scientific practice belongs to a modern epoch—a temporality inscribed into the spatial organisation of the botanical garden and *locked in* through conservation and heritage protection. The progressive linearity of modernity is playfully perverted in the story as Engler, the story's narrator, speculatively looks back on his past actions and reflects on the contemporary pressures to decolonise scientific and artistic collections and institutional knowledge. This provokes us to consider how Engler's experiences and actions laid down the conditions for the possibility of the garden changing, as he inscribed his modern progressive universalist vision of the 'world in a garden' onto what was to become the spatial layout of Berlin's Botanical Garden. With this provocation, we are made aware that the possibility of contemporary socio-ecological transformation is intimately wedded to the ongoing and historical material-discursive practices that stratify spacetimes as they interact with other regimes, such as conservation. Additionally, any chance of changing course requires knowledge of the specific historical yet contingent knowledge, practices, and regimes that lock spacetimes *in place.*

'Taxonomy Unpicked' (see above) further explores how colonial spacetimes are reproduced and the critical ruptures that break them. The story depicts the isolation and display of plants collected from Brazil with their Latin names. Over the last centuries, marked by the colonial expansion of the West, the Latin language has been established and imposed as the universal language for classifying nature. Naming is not only a descriptive practice; it gives meaning to the object considered, as embedded in a specific cosmogony. Thus, names are not universal but are local expressions of a particular form of composing the elements of the world. This can be understood as an aspect of modern spacetime, which was challenged by Alves's intervention, *You Will Go Away One Day But I Will Not*, in which she places the names of the plants as known by the Guaraní. This signified a critical juncture in the ongoing smooth spacetime of modernity and its associated epistemological claims, structuring botanical knowledge and unfolding in Berlin's Botanical Garden. While the imperative and pressure to decolonise are broader than this moment itself, the event marked a new trajectory. Modern spacetime figurations, underpinned by Western knowledge practices made possible by colonial imperialism, could no longer hold. By translocally invoking the voices, memories, and 'human-plant' knowledge and practices—that is, the lifeworlds—of the Guaraní in Berlin's garden, another spacetime began to unfurl.

Planetary Spacetime

'Palaces Diffracted' (see above) elucidates another temporality present in the structuring of the garden. It refers to a planetary time—a more-than-human geological timescale—in which the spatiality of plants across the planet is deeply entangled in geological and tectonic movements. For botanists, species associated with the 'old' and 'new' worlds are substantially different, as they have evolved on different continents with little interaction with each other since the partition of the supercontinent Pangaea (pan + Gaia) about 225–200 million years ago. In the expression 'new and old worlds', different scales of time coexist; next to the human temporality of colonial expansion, in which the discipline of botany has its roots and flourished, is geological spacetime. The 'new world' refers to the European 'discovery' of the Americas but is also considered a 'younger' world derived from the main continent of the continental partition. This geological partition and the colonial-imperial language used to describe it are reproduced in the partitioning and spatial organisation of Berlin's glasshouses, which keep 'old' and 'new' world plants firmly separated from each other, unless there is a reason to break this strict ordering to didactically present differences in the evolution of species with the same Pangaean ancestors.

This spatial bifurcation is increasingly being challenged in some botanical gardens by a different geospatial ordering system. The organisational concept of 'bioregions' divides the planet into smaller geographic units, distributed across modern nation-state boundaries, and creates translocal connections between geographically distanced territories based on biophysical and ecosystem data, alongside human settlements and cultural patterns. The reorganisation of plant geography-based on old and new worlds into bioregional spatial knowledge requires a huge effort on the part of botanical gardens to remove, store, and return plants back to the curated gardens and glasshouses. This major spatial and temporal reordering according to bioregions refers less to a colonial time or the deep geological time of Pangaea and can be understood as a critical juncture in the spatial imaginary of botanical sciences. It presents a move away from a strictly territorial spatiality and bifurcation of the world into distinct territories (old and new world) towards a more complex network of socioecological spaces, distributed across and refiguring the planet through new relationalities.

Perpetual Spacetime

Like archives, herbaria are deceptive. Herbaria seemingly present a neutral hermetically sealed container unaffected by time, but on further analysis, they protect the material foundations of how to think about the past, present, and future of botany (Foucault 1969). As such, they are considered the treasure at the heart of botanical gardens, as they accumulate botanical knowledge over decades and safeguard it in perpetuity. This is manifest in the spatial constitution of the herbarium in Berlin. Reinforcing a sense of 'spacetimelessness', the herbarium's thick concrete walls and bunker-like construction are built to be an enduring ark, tasked with storing and conserving knowledge on flora from across the planet and forever protecting against the dangers of the world.

A critical fallacy of this perpetual spacetime(lessness) can be witnessed over the last two decades, during which the impetus to digitalise herbaria seemed, at least on the surface, to turn inside out this deeply territorialised (subterranean, even) security. Arks have digital twins in digital infrastructures, which aim to store vouchers online and allow for the sharing of botanical knowledge between science and society. This means that herbaria across all botanical gardens become highly networked spaces connecting millions of plant samples and their metadata. While this seems to mark a transition from a territorial bunker-like spatial logic to an open-access nonhierarchical distribution of knowledge, or a network space, on closer analysis, we see some of the territorialising boundaries, hierarchies, and privileges translate across digital borders, producing new hierarchies.

Polycontexturality

While the design of the Botaniverse aimed to reassemble new stories from existing data, it was also curated with visitors in mind, in a sense following the constraints and obligations of the design of botanical gardens outside the digital world. The curation followed several principles: 1) delineating the space of each story as a territory, whose boundaries are enacted through the sound of the voiceover of each narrator (known as spatial audio), drawing an invisible circle around the story spot and allowing for locating sound islands in the digital space; 2) organising visitor's circulation through the Botaniverse by making it possible for visitors to navigate by sight (by placing titles above each story island and creating teleportation portals between the stories); and 3) allowing for a variety of experiences of the stories, for example, by distinguishing between the garden landscape level and a subterranean level of networked caves. In contrast to the highly interpreted landscapes above, the cave spaces staged raw data without interpretation, such as interview recordings that require longer reception time and space for visitors' own inspirations.

This brings another temporality into the mix, that of the visitor's experience. As multiple avatars simultaneously circulate through the Botaniverse, their real bodies occupy desks fixed in different spacetimes across the world. Visitors weave through the spatial sequencing of stories staged across the digital garden and paced by the rhythm of intersecting narrations and animations as the spatial audio technology kicks in (Figure 5). As they move over invisible hotspots and teleportals, their digital twins accelerate, slow down, or wrap through digitised botanical space (Figure 6, 7). Polycontexturality accounts for being in different spaces at the same time (Löw and Knoblauch 2022). It highlights the experience of being in two (or more) places at once, like being on a Zoom call. In this way, the subject has to make sense of the different spatial logics associated with, for example, being on a phone in a cafe while at the same time operating in a digitalised space in which other participants are evoking their multiple spatialities within the interaction. The visitor's experience of the Botaniverse has a similar constitution, but the experience has become more complex, as multiple temporalities overlap and interfere with one another.

Conclusion: Spacetime Matters

Curating the Botaniverse allowed not only for a deeper analysis of spacetime entanglements within our botanical garden research but also for a multimodal analysis and novel form of engaging a wider audience beyond the usual academic dissemination routes. Provoked by the desire to overcome the stasis of mapping, we stepped into the map space, mistaking it for territory, and, by doing so, our mapping practices were expanded to include multiple formats and modalities, such as videos, animations, drawings, texts, audio recordings, photographs, voiceover, and so on. Through the experiment of the Botaniverse, we created an immersive hybrid spatial experience in which data can be collaboratively analysed remotely while stories about space and time are displayed for other public visitors to engage with. It allows multiple times and spaces to be grasped together. Echoing Massey's (2001) theoretical provocation, as described above, immersive mapping presents a horizon for considering how we can conduct methodological research and analyse the multiple ways in which time and space are entangled in empirical cases. For the spatial researcher, this means questioning not only the ways in which space is produced or constituted in practice but also paying close attention to how time is produced or constituted in practice—that is, viewing not only how spatial orders are fixed *in* time or *over* time but also how spatial orders structure time and, conversely, how temporal orders structure space. These are complex challenges, both analytically and methodologically, and require much more working out in the future. However, it is our conviction that, in order to take the implications of refiguration seriously—that is, to go beyond the modern programme of scalar and linear globalisation and think dialectically of counter processes, opposing social force, and resistances—it is necessary to find new ways to grasp spacetimes in their multiplicities. We hope that immersive mapping will offer a foothold for further spacetime experiments.

Funding Statement
The research was funded by the German Research Foundation (Deutsche Forschungsgemeinschaft). Projektnummer 290045248 – SFB 1265.

Image Credits
Figure 1: Hybrid mapping by Julius Frittkau & Zoe Hochstein-Morran, realised within the MA-course 'Hybrid Space, Hybrid Knowledge' at the Technische Universität Berlin, Summer Semester 2021
Figure 2, 3, 4, 5, 6, 7: Jamie-Scott Baxter, Lus Constantin, Séverine Marguin with the help of Mara Oppermann

References
Adams, William. 2013. *Against Extinction: The Story of Conservation*. London: Earthscan.
Baber, Zaheer. 2016. 'The Plants of Empire: Botanic Gardens, Colonial Power and Botanical Knowledge'. *Journal of Contemporary Asia* 46(4): pp. 659–679. https://doi.org/10.1080/00472336.2016.1185796.
Barad, Karen. 2007. *Meeting the Universe Halfway. Quantum Physics and the Entanglement of Matter and Meaning*. Durham: Duke University Press. https://doi.org/10.1215/9780822388128.
Baxter, Jamie S. 2023. *Self-Organised/Ing Infrastructures. Understanding Everyday Practices of Spread and Resistance in Processes of Socio-Spatial Change on the Periphery. Multisited Ethnographies in Rural Portugal and Austria*. PhD Thesis, Technische Universität Berlin.
Baxter, Jamie S., and Séverine Marguin. 2024. 'The Refiguration of Conservation: Introducing the Concept of "Staging Nature" in the Case of Botanical Gardens'. *Museum and Society* 22(1): pp. 14–33.
Baxter, Jamie S., and Séverine Marguin. 2025. 'The Mimetic Qualities of Space: Staging Nature at Botanical Gardens'. In *The Social Quality of Public Space. Integration, Strategy, Subjectivation*, edited by Letteria G. Fassari and Martina Löw. London: Routledge.
Baxter, Jamie S., Séverine Marguin, Sophie Mélix, Martin Schinagl, Ajit Singh, and Vivien Sommer. 2024. 'Hybrid Mapping: A Methodology to Study Relational Space. The Case of Botanical Gardens Berlin'. *Architecture and Culture* 11(3).
Bennett, Jane. 2010. *Vibrant Matter: A Political Ecology of Things*. Durham, NC: Duke University Press.
Braverman, Irus. 2023. *Settling Nature: The Conservation Regime in Palestine-Israel*. Minneapolis: University of Minnesota Press.
Chakrabarty, Dipesh. 2000. *Provincializing Europe: Postcolonial Thought and Historical*. Princeton: Princeton University Press.
Chakrabarty, Dipesh. 2018. 'Anthropocene Time'. *History and Theory* 57(1): pp. 5–32. https://doi.org/10.1111/hith.12044.
Coole, Diana, and Samantha Frost, eds. 2010. *New Materialisms. Ontology, Agency, and Politics*. Durham, NC: Duke University Press.
Descola, Philippe. 2005. *Par-Delà Nature et Culture*. Bibliothèque Des Sciences Humaines. Paris: Gallimard.
Ernstson, Henrik, and Erik Swyngedouw, eds. 2019. *Urban Political Ecology in the Anthropo-Obscene Interruptions and Possibilities*. London: Routledge.
Fabian, Johannes. 1983. *Time and the Other: How Anthropology Makes Its Object*. New York: Columbia University Press.
Foucault, Michel. 1969. *L'archéologie Du Savoir*. Paris: Gallimard.
Gandy, Matthew. 2024. *Natura Urbana Ecological Constellations in Urban Space*. Cambridge: MIT Press.
Haraway, Donna. 1991. 'A Cyborg Manifesto: Science, Technology, and Socialist-Feminism in the Late Twentieth Century'. In *Simians, Cyborgs and Women: The Reinvention of Nature*, pp. 149–181. New York: Routledge.
Jameson, Fredric. 2003. 'The End of Temporality'. *Critical Inquiry* 29(4): pp. 695–718. https://doi.org/10.1086/377726.
Kaiser, Katja. 2022. 'Duplicate Networks: The Berlin Botanical Institutions as a "Clearing House" for Colonial Plant Material, 1891–1920'. *The British Journal for the History of Science* 55(3): pp. 279–296. https://doi.org/10.1017/S0007087422000139.
Knoblauch, Hubert. 2022. 'Multiple Säkularität, Multiple Räumlichkeit Oder Multipolarität – Auch Eine Kritik Der Postkolonialen Vernunft'. In *Verstehen Als Zugang Zur Welt. Soziologische Perspektiven*, edited by Uta Karstein, Marian Burchardt, and Thomas Schmidt-Lux, pp. 99–118. Frankfurt Main: Campus Verlag.
Knoblauch, Hubert, and Martina Löw. 2024. 'The Refiguration of the Global: Globalization and the Spatial Logics of Digitalization'. *Critical Sociology* (online first). https://doi.org/10.1177/08969205241262304.
Korzybski, Alfred. 1933. 'A non-aristotelian system and its necessity for rigour in mathematics and physics'. In *Science and Sanity: An introduction to non-aristotelian systems and general semantics*, edited by Alfred Korzybski, pp. 747–761. Lakeville: International Non-Aristotelian Library.
Koselleck, Reinhart. 2004. *Futures Past. On the Semantics of Historical Time*. New York: Columbia University Press.
Koselleck, Reinhart. 2018. *Sediments of Time On Possible Histories*. Redwood City: Stanford University Press.
Latour, Bruno. 1993. *We Have Never Been Modern*. Cambridge: Harvard University Press.
Löw, Martina, and Hubert Knoblauch. 2022. 'The Refiguration of Space, Circulation, and Mobility'. In *Spatial Transformations. Kaleidoscopic Perspectives on the Refiguration of Space*, edited by Angela Million, Christian Haid, Ignacio Castillo Ulloa, and Nina Baur, pp. 17–27. London: Routledge.
Marguin, Séverine, Dagmar Pelger, and Jörg Stollmann. 2024. 'Mappings as Joint Spatial Display'. In *Handbook of Qualitative and Visual Methods in Spatial Research*, edited by Anna Juliane Heinrich, Séverine Marguin, Angela Million, and Jörg Stollmann, pp. 295–311. Bielefeld: transcript Verlag.
Massey, Doreen. 2001. 'Talking of Space-time'. *Transactions of the Institute of British Geographers* 26(2): pp. 257–261. https://doi.org/10.1111/1475-5661.00019.
Massey, Doreen. 2005. *For Space*. London: SAGE Publications.
Mudimbe, Valentin Y. 1988. *The Invention of Africa: Gnosis, Philosophy, and the Order of Knowledge (African Systems of Thought)*. Bloomington: Indiana University Press.
Tsing, Anna L., Andrew S. Mathews, and Nils Bubandt. 2019. 'Patchy Anthropocene: Landscape Structure, Multispecies History, and the Retooling of Anthropology'. *Current Anthropology* 60(S20): pp. 186–1997.
Weidenhaus, Gunter. 2017. 'Der Zusammenhang von Raum und Zeit: Zur Konzeption sozialer Raumzeit und ihrer empirischen Relevanz in Biografien'. In *Zeitschrift für theoretische Soziologie. 4. Sonderband. Raum und Zeit. Soziologische Beobachtungen zur gesellschaftlichen Raumzeit*, edited by Henning Laux, Anna Henkel, and Fabian Anicker, pp. 40–73. Weinheim: Beltz Juventa.

Long-Now Mapping

Noël van Dooren

A Landscape View of How to Represent Future Dynamic Systems

Introduction

Landscape is a living system and therefore subject to change and dynamism in many ways. Geological, climatological, and biological processes are at work, form vulcanism to tree growth, and cross with changing societal needs, from land use alteration to migration. As a landscape architect, this engages me. When a designer intervenes in landscape—be it making a new park, restructuring a river, or envisioning urban extensions—drawings are produced to explore, discuss, and communicate the design. It seems evident that drawings of designed landscapes explore and reflect the dynamism and change over time that is so characteristic of a landscape. Therefore, both a question and a statement guide this essay. The statement is that space and time *should* be held together. The preceding question is*: how can space and time be held together in a drawing?* To explore this, I will dive into the history and nature of design drawings, point to the influential character of professional practice, and propose a shared idea of landscape. Central to this essay is an overview of the ways of understanding the dimensions of time in landscape. From that, I move into how this issue was and is applied in design practice, and end by positioning the issue in today's debate.

Context

There is an incredible richness in drawings—both the types of drawings, and the roles of drawings— even if we restrict the area to design drawings. To answer how time and space can be held together in drawings, one important preliminary observation is that landscape design drawings fit into the repertoire of drawings that primarily stem from the history of architecture—that is, the orthogonal plan, section, and view (see, e.g., Blau and Kaufman 1989). Such drawings form the basic ingredients of new building proposals and, as translated into landscape, new landscapes. However, in design processes, we see a distinction in a set of observational and analytical drawings, in documenting and researching the necessary information to arrive at a design, and in the exploratory drawings that show future situations—the design.

When it comes to space and time, there is quite a large body of observational and analytical representations of how these two relate—see, for example, the beautiful collection in Tufte's (1990) *Envisioning Information*. The last chapter of Tufte's book is appropriately titled 'Narratives of Time and Space', as it shows, for example, dance instructions of how to move in space next to railroad timetables and water supply schemes all spread out in space. Different from such observational drawings, drawings in the design field are often made to explore, compose, build, and manage future situations. It is particularly in such drawings that designers cope with unknown future situations. Many of the drawings in Tufte's collections are diagrams or maps. Maps as objects or mapping as a process in a designer's approach can be extremely speculative and future oriented.

Figure 1, a map of Greenland titled 'Nomads of the New North', explores a possible future in which a 5°C rise in temperature makes large parts of the world inhabitable. The narrative proposes a world in which land ownership and borders are history, and where people follow the climate in a seasonal pattern. In fact, this map does not specifically represent time. It is situated somewhere in the twenty-second century, and one could say that it snapshots an episode in the expected turbulent future. 'Dating' a drawing is an important first step to locating it in time, but in speculative research like this, exact dating would not be appropriate. This future-oriented domain is the area that I am specifically looking at. In this regard, the principal distinction between observational and explorative drawings is crucial. Designers may study the dynamism at hand in their drawings using an observational approach, but primarily it is *intended* change or *expected* dynamism that is shown, which calls for an explorative approach.

To position such drawings, it is important to note that a major part of the designer's work is organised into *projects*. The designer is asked to do

Figure 1 Humanity on the Move—Beyond Borders Greenland speculative map (Lena Knappers and Bram van Ooijen, 2023). What if a 5°C rise in temperature makes large parts of the world inhabitable and people follow the climate in a seasonal pattern? The map is situated somewhere in the twenty-second century—it is a snapshot of an episode in the expected turbulent future.

something in a certain time frame at a well-defined location, commissioned in terms of a contract. The project ends with a product that may be a report with a written explanation and a package of drawings. Often, final products consist of at least twenty to thirty drawings, neglecting the numerous sketches and drafts made during the design process. Intriguingly, drawings have their own audience and exist in their own right; see, for example, Goffi's (2007) concept of the *twinned body.* Goffi argues that the drawing has its own existence next to the realised project, and the project may develop and even disappear, while drawings can outlive it (and stay unaltered). Drawings are embedded with knowledge, history, and cultural values, just as they can be understood as a process—one of ideation, exploration, and communication (van Dooren 2010).

However, in most cases, drawings implicitly or explicitly communicate with clients, stakeholders, society, and colleagues, as projects are executed in 'real-world' contexts. Money, time, and power are involved, as are broader societal and cultural notions that co-shape how professional practice works (think of why certain ways of drawing are in vogue today or not). In *The Reflective Practitioner*, Schön (1983) provides a seminal contribution regarding how professionals operate in this 'muddy' context. This also concerns design drawing; for example, the idea that drawings should hold together space and time is, in practice, confronted with how clients, design firms, and public bodies operate in an interdependent way within a larger socio-economic system. In this context, what *should* be done is very different from what *is* done in actuality. In relation to the overall goal of this essay, this is obviously crucial, as, in this context of muddy practice, studying time and space together in drawings apparently is not self-evident. In practice, landscape architects are not so eager to follow up discoveries concerning the notation of time (van Dooren 2017). Even when many designers support that drawings contain information on space and time, in a project context the focus may be on what has to be delivered to make the project happen. A rather implicit, diverse, and practical reasoning is at work. Most landscape architects consider expertise on the relation between time and space as an evident part of their knowledge and work as supposedly evident. However, there is no urgency felt to draw time. It is not asked for, and as drawing time potentially complicates smooth design processes, there may even be reasons *not* to do it—a typical example of 'muddy practice'.

Landscape

This essay started with a description of landscape as a living system, and from there, a question developed about time and space in drawings, with drawings to be made in case interventions in landscape are prepared. Thus, landscape is a crucial notion here and has a broad meaning. The European Landscape Convention speaks of landscape as 'an area as perceived by people whose character is the result of the action and interaction of natural and/or human factors'. This applies to natural, urban, and peri-urban areas and to beautiful or everyday or degraded areas, from industry to leisure areas, and from flood plain to inner city. All these landscapes can be part of design processes to change them for any sort of reason.

Landscape is a physical category, but it is also a system in which processes are at work—ebb and flood, the seasons, erosion, evolution. In the human and urban landscape, mobility, work, and leisure, to mention a few, lead to ongoing change and dynamism. There is an intensive dialogue between the understanding of landscape on the one hand and design approaches on the other. Consequently, drawings are pivotal in this dialogue. They are the metaphorical space for observing, explaining, and exploring this. Landscape architects, in general, feel that it is an evident part of their practice to cope with the dynamism of landscape or, put more strongly, to look at it as an impetus for design. Figure 2 is an example of this. The drawing presents a typical landscape architecture narrative—in this case, for Dutch polders. It is a four-part

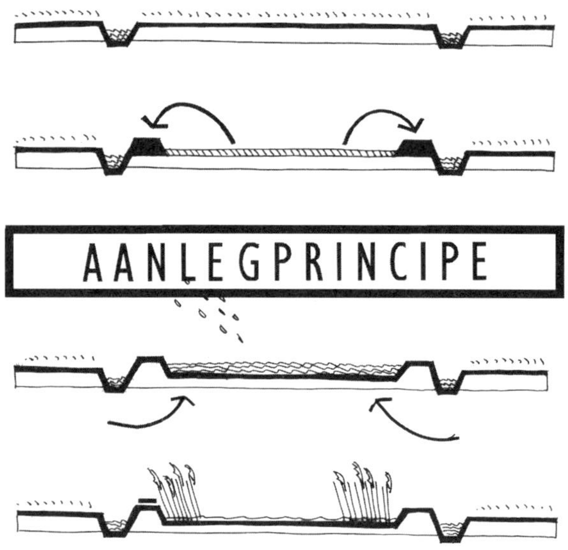

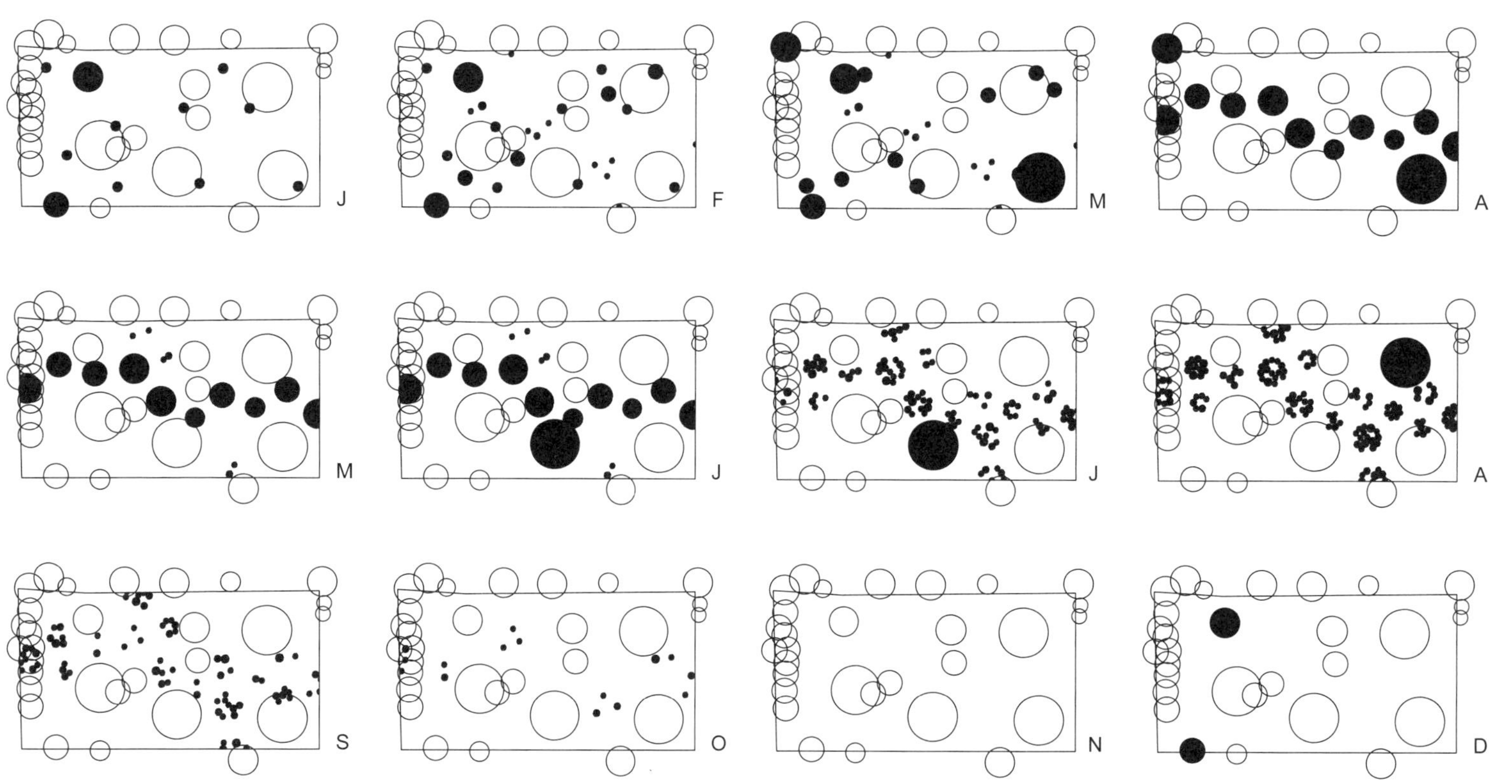

Figure 2 Sectional diagram showing four points in time (*Plans Without Final Image*) (Buro Lubbers, 1998). The drawing is a four-part narrative that reveals how it is today (1), followed by a controlled intervention (2), causing rainfall to create a rather permanent water body kept by the small dikes (3) and thus inviting natural processes to build a new landscape (4).
Figure 3 Lace Garden, Amsterdam, plan diagram (Anouk Vogel, 2009). This garden design drawing shows seasonal performance. The flowering is displayed as 'yes' or 'no'. The drawing is divided into months, as this garden was designed to be relevant throughout the year.

narrative that first reveals how it is today, followed by a controlled intervention of digging and shaping the earth into small dikes. The third part shows the inevitable consequence of the intervention—rainfall leads to a rather permanent water body kept by the small dikes. This consequence of the intervention is intended (see the last episode), as it will 'invite' natural processes to build a new landscape. The time frame is unspecified but could be something like 'the next season'. The importance of this example is that drawing time is not only instrumental in explaining or expanding on the issue. Here, drawing time is inextricably interwoven with a design approach that sees dynamism and natural forces as essential partners in building the project. Landscape design in that approach is not directed towards a neat final situation but towards creating basic conditions that give way to and invite natural forces. This implies substantial knowledge of how these processes work, as they are the co-constructors of what should be reached over time.

The last example may have given quite a straightforward answer to the question of how to work with space and time in a drawing. However, professional reality complicates matters. Even if the statement that space and time should be held together is approved, and even if it is known how to accomplish this, an eager application in practice is something different. Working with time is felt as an evident part of practice, but it is also experienced as a difficult task in terms of informing both the client and the audience, as they are often not experts in such dynamism. Drawings are a vehicle for this exchange of information, but they operate in a narrow strategic field—namely on how to avoid complicating matters and endangering the decision-making process. The fact that drawings in landscape architecture are strongly rooted in the architectural tradition is both a blessing and an obstacle. It obviously helps that landscape architects can navigate practice with the representational findings as invented in architecture. However, architectural representation strongly stresses the outcome of design processes as ready and final products. Even if one can indeed point to a specific moment of readiness (the opening, the party, etc.), the process of using it, changing it, or fighting the weathering process leaves this one moment of readiness questionable (Leatherbarrow 2009). Still, this idea of readiness deeply influences how drawings are read. In fact, one can say that landscape architecture not only had (and has) to solve specific disciplinary issues—for example, how to notate landscape elements such as a swamp in a legend (De Jong et al. 2008)—but also has to aim for a new reading of drawings in which risk, uncertainty, open-endedness, and tendency are accepted.

Understanding Time in Landscape

The history, understanding, and representation of time is a huge topic that crosses disciplines and domains, from physics to philosophy, and the arts. Here, I focus on a hands-on perspective on the specific nature of time in landscape. I primarily follow a very fitting distinction in timescales, as proposed by the Long Now Foundation ('encouraging imagination at the timescale of civilisation', as longnow.org states). Landscape in this approach is positioned in time on three fundamentally different scales, as represented in a beautiful graph by Long Now Foundation (n.d.).

The first is the scale of 'these days', for example, the flowering of a garden or the winter season. Figure 3 explores this time frame as part of a garden design. The drawing shows seasonal performance. The diagram displays the flowering of a garden, not in the actual colours but rather as reduced to 'yes' or 'no', thus underlining the aesthetic considerations in making a drawing and the choices to be made with regard to the informational value. Note that the drawing is divided into months. Often, flowering is seen as a spring and summer phenomenon. As this garden was designed to be relevant throughout the year, a monthly scheme was seen as the appropriate rhythm.

The second timescale is 'our time', meaning that what we can experience in our lifetime or generation—say ten to a hundred years. Figure 4 is a typical

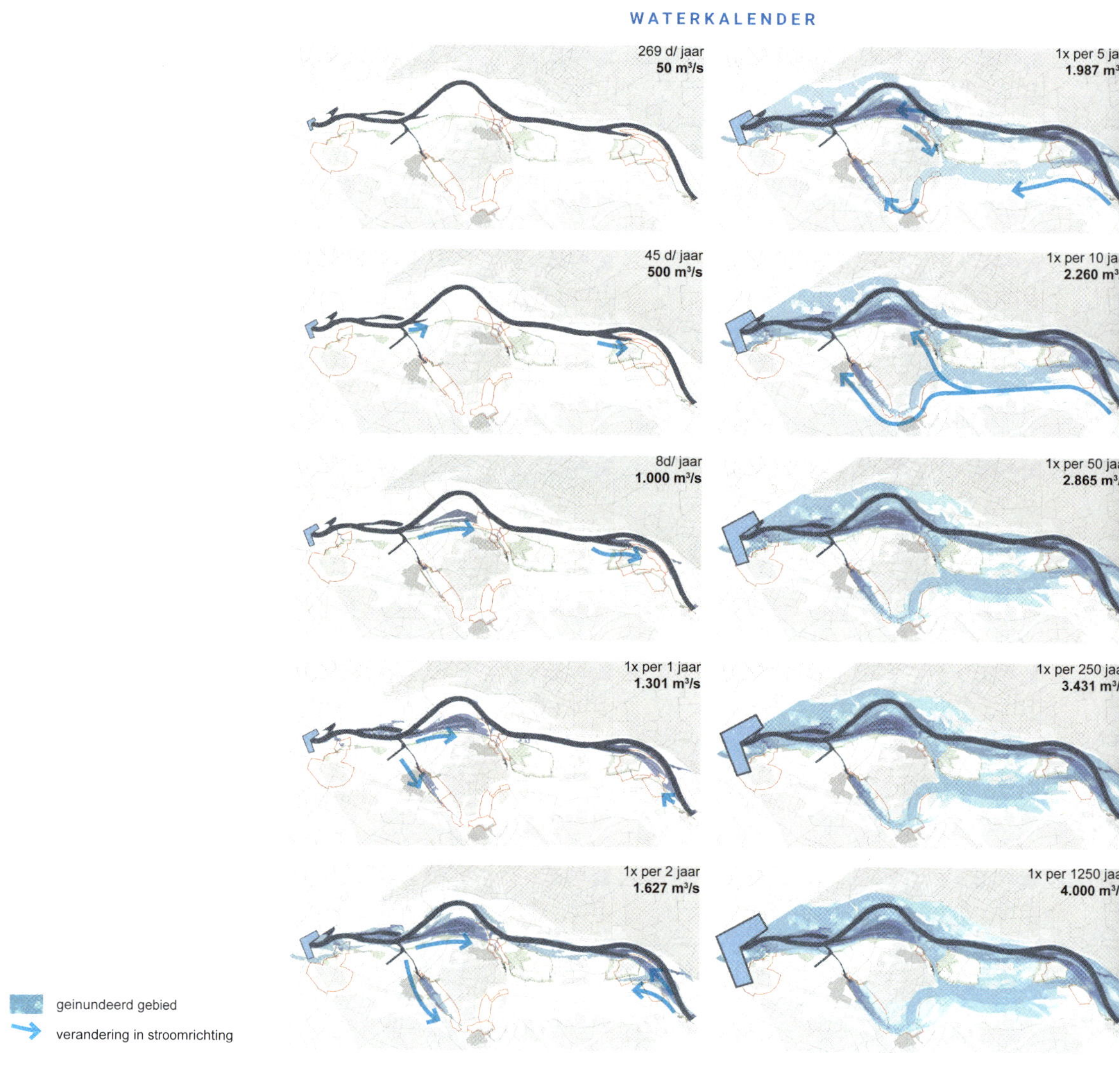

Figure 4 Water calendar, Ooijen-Wanssum river Maas reconstruction (H+N+S landscape architects, 2019). In this 'map diagram', we see how a river operates in ten relevant happenings, ranging from 'a few times per year' to 'once in 1250 years', with the latter referring to Dutch safety norms. The map combines three stages in one (here, the one-, two-, and ten-year intervals).

expression of this. The designers indicate this drawing as a 'water calendar'. In this 'map diagram', we can see how a river operates in ten relevant happenings, ranging from 'a few times per year' to 'once in 1250 years', with the latter referring to Dutch safety norms. It is interesting to compare this water calendar that displays the events as separate stages with a map that combines three stages in one (here, the one-, two-, and ten-year intervals). Note that a ten-year interval, or, for that matter, a twenty- or fifty-year interval, is experienced as exceptional, but with a large probability, this time frame will happen in our lifetime. In this case, drawing time can be discussed in three contributions to the design process. First, it informs the designers. Knowing how these systems work in general, drawings can show the designer how they should work in this specific situation. Second, the drawing act is a virtual test run of the designed intervention in dialogue with the technicians (e.g., did we reach our goals?). Last but not least, it is part of the public exchange to attain involvement with and understanding of the proposed landscape.

The third timescale is the 'long now'—ice ages, geological processes—in short, millennia. Even if that timescale is far too slow for our human understanding, major parts of our surroundings are shaped by processes operating on this scale, and they keep doing so, whether we are aware or not. Figure 5 is part of a research project in which the dynamism of 'nonhuman actants', such as erosion and therefore falling stones in the Norwegian rural landscape, is mapped. Erosion, rain, ice, and wind are continuous processes, but they have become stronger and less predictable. The project studies how to engage with risk in this rural landscape due to climate change. The map is a fascinating combination of factual information produced by LIDAR laser scans and a drawing process in which the erosion forces are more or less reenacted with pencil on paper. The research project seeks to unravel this by drawing the dynamics and crafts that shape the landscape. The LIDAR basic information describes the landscape 'as it is' for an exact specified time frame. Mapping dynamism implies an understanding of the forces at work at the scale of the watershed. Building upon that, this could lead towards designed interventions. The same drawing process as that used to depict the actual situation could then become an *ex ante* evaluation of how such interventions would keep the risks of erosion at bay.

Surprisingly, this organisation of time in landscape in three very different timescales is not an approach often mentioned. By far, the most popular categorisation is in circular and progressive time; for example, the seasons are a returning pattern of change over the year and therefore represent circular time, whereas a tree grows on and on, illustrating progressive time (e.g., Zerubavel 2003). In my view, there is no tension between these views; a distinction in circular and progressive time involves adding up to and enriching an understanding of timescales. Think, for example, of a tree, yearly going through a seasonal pattern (these days), slowly growing monumental (our time), and in a very wide perspective, transferring into carbon (the long now).

In a landscape context, these three time scales are very relevant to understanding the quick and slow pace of change. They can be looked at separately, but they also operate in parallel and, to some extent, interact. In this interaction, the concept of uncertainty is important. Think, for example, of vulcanism as a typical phenomenon related to the long-time scale of geological processes. It happens rarely, but it certainly will happen from time to time, and when it happens, it strongly influences the actual seasonal performance or the course of events in the coming decades. Reading the future of landscapes in terms of events, disasters, and happenings suggests a more narrational understanding of time, and from a design perspective, it asks for anticipation of the uncertainty and risks.

A Design Perspective: Repton and Halprin

If we move to a design perspective on time in landscape, the narrational view of 'before and after' is crucial. In a design view, this very simple model of

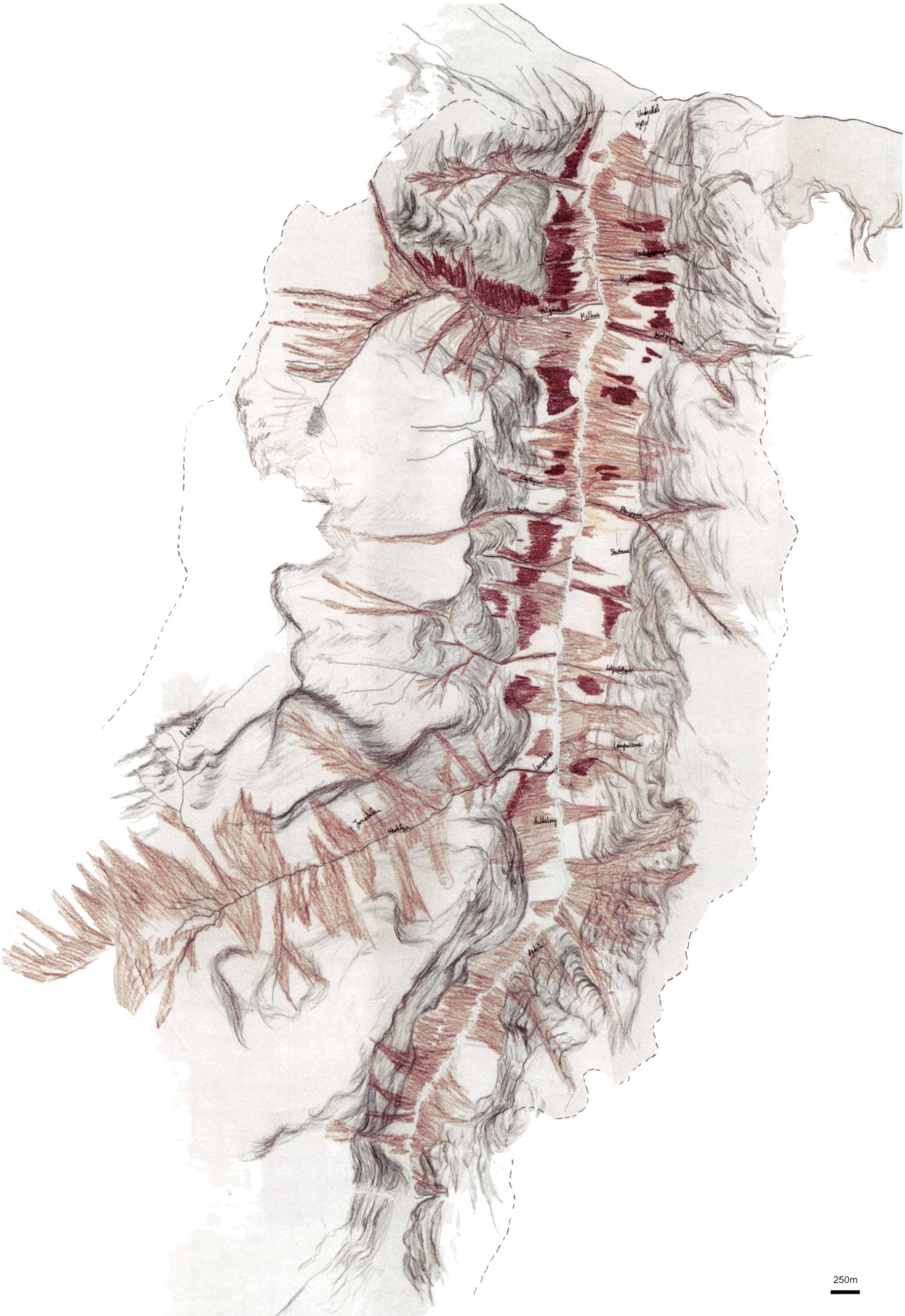

Figure 5 What Slides, Crumbles, Drifts. Situation Spring 2022, interpretative map (Violaine Forsberg Mussault, 2023). Here the dynamism of 'nonhuman actants', such as erosion and therefore falling stones in the Norwegian rural landscape, is mapped. The project studies how to engage with risk in this rural landscape due to climate change.

time relates not so much to the nature of landscape as to some sort of action. Obviously, natural forces can be at work, but in a design context, 'before a nd after' refers to the very idea of making and implementing a design—or to intervening in landscape to change it. Here, we arrive at the history of representing time in landscape design.

In the history of art, cartography, and information design, to mention a few, the representation of time has been a quest throughout the ages, as can be seen in, for example, *Cartographies of Time* (Rosenberg and Grafton 2010). In landscape architecture, the first explicit attempt was made by English 'landscape gardener' Humphry Repton in the late eighteenth century (Loudon 1988). With the 'before–after' drawings he proposed, he invented a representational technique to realise that. Repton inserted in these before–after drawings a part that could be flipped. In position A, it showed the landscape as it was; in position B, one could see the landscape as desired. Obviously, this is a very basic view of time and dynamism, but as a chapter in 'representational inventions', it is crucial.

The further development of drawing time in landscape architecture is a hop, a step, and a jump in which individual characters contributed for their own reasons. One example is Lawrence Halprin, a main figure in American landscape architecture in the nineteen-sixties and nineteen-seventies. Halprin was married to Ann Halprin, a choreographer, and became interested in how drawings could deal with 'movement in space' (Halprin 1969; Hirsch 2011). I use the word 'deal with', as Halprin looked at it in two ways. One involved how drawings could *register* movement in space. He explored notational techniques to do so, for example, in his study of how people moved in a street. The second involved how drawings could *instruct* movement in space. Halprin made several 'design' drawings that, as choreographies, almost inform people how to move and act over time and in space.

Inspired by Halprin, I staged workshops with choreographers to exchange ways of drawing. Figure 6 shows how Belgian choreographer Louis Vanhaverbeke—a student at the time the drawing was made—used drawing as a design means to develop a choreography. The drawing can be seen not only as an instruction of how and when to act but also as a speculative registration of what will happen on the stage.

With his experimental drawing, Halprin pushed the boundaries of the repertoire of drawings in use in landscape design processes. Following his inspiration from choreography, he introduced the *score* into landscape architecture as a useful addition to that repertoire. The score, particularly in relation to dance, instructs on *who* is in charge, *what* is supposed to be done, *when* to do it, and *where* to do it. Halprin's quest was how to translate this into landscape architecture drawing. Again, just as in the case of Repton, this was a revolutionary step in thinking about representation.

Despite Halprin, the known repertoire of landscape architecture up to today tends to prioritise *what* is done *where* (van Dooren 2017). In my view, drawings in which space and time are held together are capable of showing in one drawing not only the 'what and where' but also the 'who and when'—the 'who' may refer to a person, but also to a force in nature. Scores, as Halprin proposed, have the quality to do this. Combining all this information into one drawing can challenge the limits. Halprin used scores in performance-like situations. Once they are applied to real-world design problems, scores can become very complex, as can be seen in the score (classified as such by me) from the landscape architecture office GROSS.MAX in their competition entry for Tempelhof, Berlin (Figure 7). In this score, we see a rough idea of which parties are expected to act in order to realise the intended development. Note that a connection to topography is absent here. This relation to *where* would make the drawing even more complex.

Figure 6 Untitled drawing as part of performance score (Louis Vanhaverbeke, 2013)

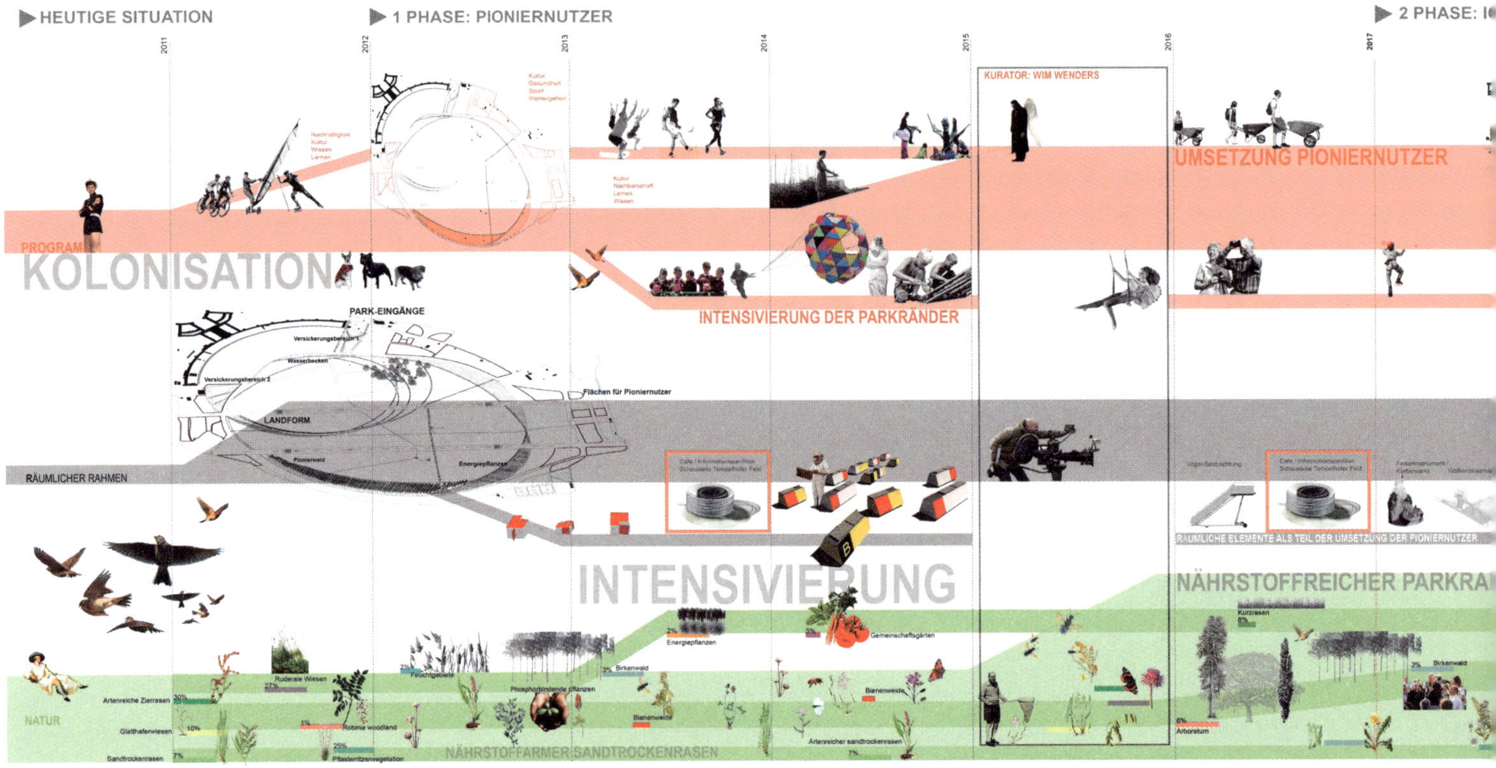

Figure 7 Parklandschaft Tempelhof, timeline or score (GROSS.MAX, 2011). This score, part of the competition entry for Tempelhof, Berlin, describes which parties, over a time frame of decades, are expected to act in order to realise the intended development.

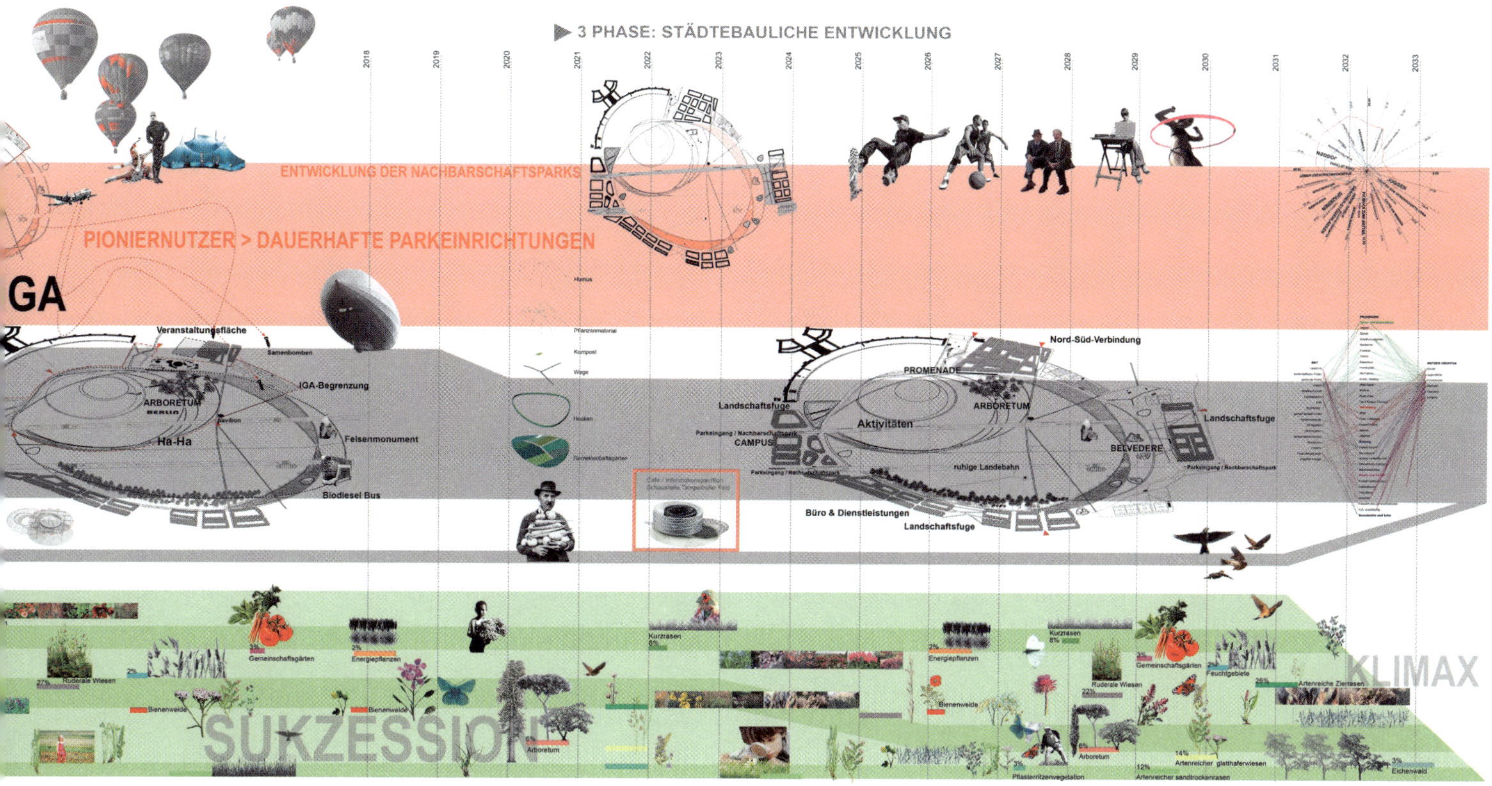

Landscape Urbanism

If Repton and Halprin, among others, made individual contributions to a slowly emerging discourse on time, representation, and landscape architecture, a group of designers and theoreticians under the umbrella of *landscape urbanism* provided a solid basis. Corner's (1992) essay 'Representation and Landscape' is a crucial step forward. Corner (1992, 146) notes the fundamental issue that 'three unique qualities [the spatiality, temporality, and materiality of landscape] evade reproduction in other art forms and pose the greatest difficulty for landscape architectural drawing'. Corner but also Waldheim, editor of *The Landscape Urbanism Reader* (Waldheim 2006), which discusses the relation between landscape, dynamics, and time in several essays, were in fact the first to verbalise this discourse and locate it in the wider field of theory, teaching, and practice. Thereby they also stimulated a new approach towards landscape. Landscape became seen as an agency, or a performative entity. In his seminal reader, Waldheim (2006, 39) states, 'Landscape is a medium ... uniquely capable of responding to temporal change, transformation, adaptation, and succession. These qualities recommend landscape as an analogy to contemporary processes of urbanisation and as a medium uniquely suited to the open-endedness, indeterminacy, and change demanded by contemporary urban conditions.' With this, the frontrunners of landscape urbanism also opened up new roads for landscape architecture—that is, claiming a role in the large global and societal problems of these days, in which uncertainty, change, and open-endedness are key.

Notably, the discourse in *The Landscape Urbanism Reader* (Waldheim 2006) is first and foremost about landscape as a medium. Later, a shift can be seen towards landscape architecture and design practice in a monograph on Michel Desvigne by Corner: 'Desvigne views landscape architecture as a work in process, never really attaining an ideal state at any moment in time, but always exceeding expectations when set in motion over time, when viewed as an active palimpsest accruing new properties, qualities, and potentials in

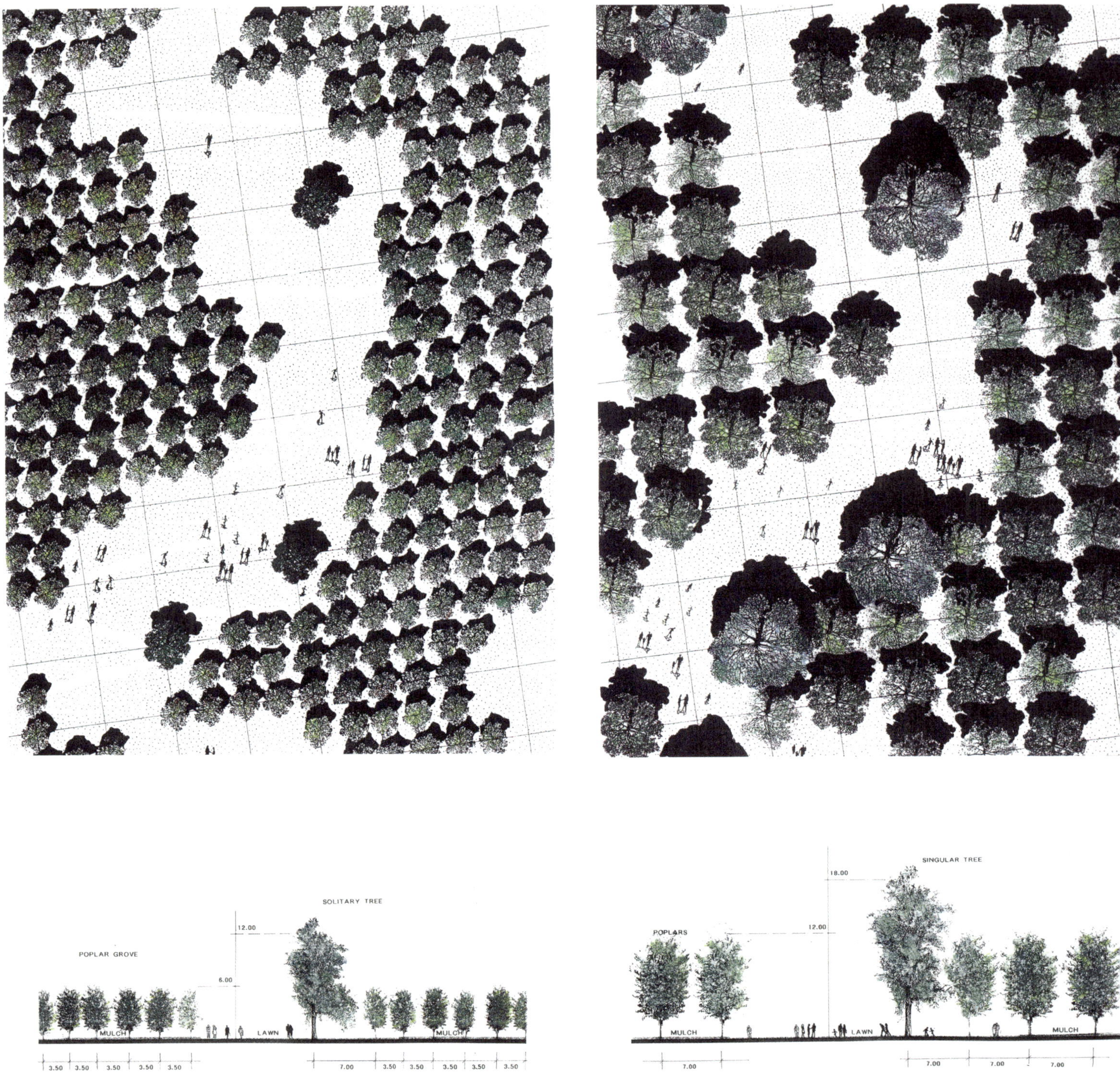

Figure 8 Greenwich Millennium Park, London, mixed drawing (plan and section showing four points in time) (MDP Michel Desvigne paysagiste, 2000). Desvigne displays the evolution of the project with four moments in time. The implicit statement is that a design for a green public space cannot be represented in terms of one final state, as it never attains an ideal state at any moment in time.

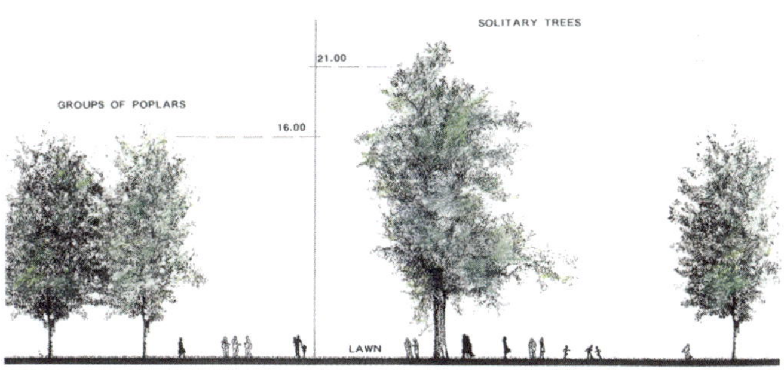
SOLITARY TREES
21.00
GROUPS OF POPLARS
16.00
LAWN

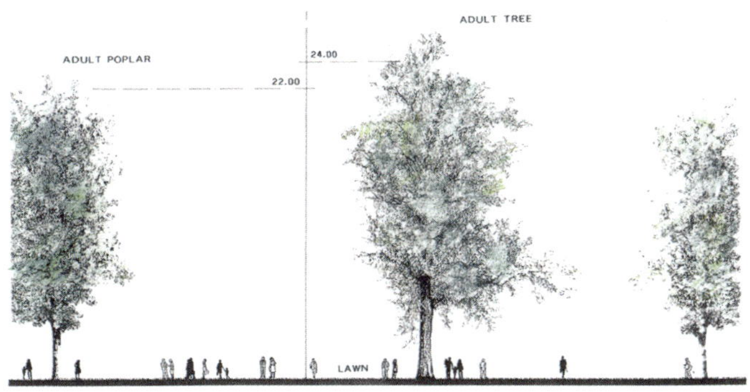
ADULT TREE
24.00
ADULT POPLAR
22.00
LAWN

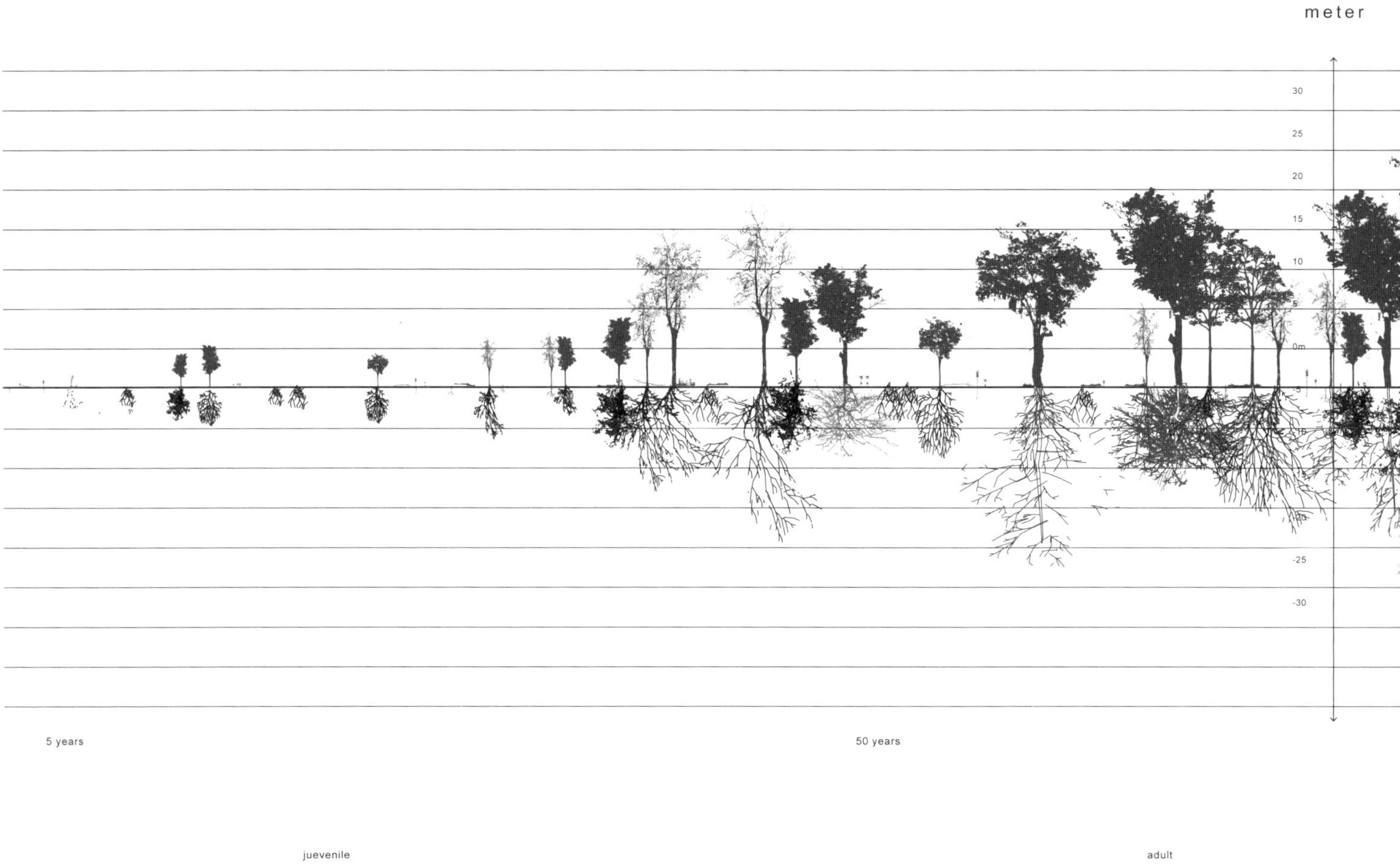

Figure 9 Growing Horizons. Investigations of Soil and the Underground. The Proposal of a New Landscape Typology in Southwestern Denmark, ‘Projected Root/Crown Growth 5–100 Years’ (Johanna Bendlin, 2020). The drawing relates to an afforestation study project. Here, forests are planted as a design act and as part of a larger transition. The afforestation is done in the low-lying southern Denmark meltwater floodplains, which are threatened by rising sea levels.

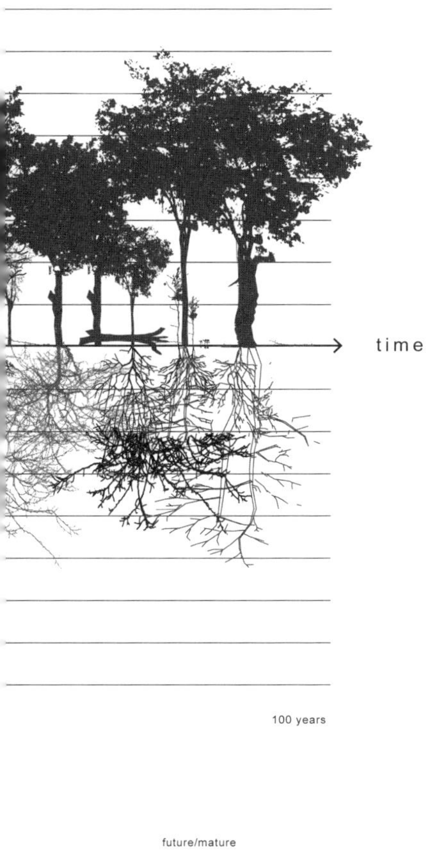

time.' (Corner 2009, 9). Figure 8 illustrates this very well. In this 'mixed' drawing (which combines both a plan and section in a series of four time points) for Millennium Park, realised in London in 2000, Desvigne displays the evolution of the project with four moments in time. As the moments are not specified, this can be considered an abstraction of the evolution. I see this drawing as an iconic example due to its implicit statement that a design for a green public space cannot be represented in terms of one final state, as it never attains an ideal state at any moment in time.

The contributions of Corner, Waldheim, and the like strongly helped institutionalise drawing time—in the first place, as a part of teaching curricula. For example, Harvard started to post on its website about landscape urbanism course items in which a range from 'mapping ecological systems to illustrating time-based processes' was offered (van Dooren 2017, 78). As curricula in landscape architecture worldwide have adopted this approach in terms of what landscape architecture is and how drawings can be made, drawings displaying time aspects have become an increasingly known category.

A Manual for Drawing Time and Space

How to draw orthogonal plans, sections, and views has been long known. The same can be said for how to draw visualisations, which are very popular today. Most of these drawings are not dated; instead, they suggest some sort of 'ready state' to be reached and kept in a linear process. In the case of landscape, this is a fundamental flaw (van Dooren 2018). There is no final or ready state, or it would be the moment that we declare a project to have arrived at a 'normal' landscape, which is subject to change, as all non-designed landscapes in general are. Figure 9, which depicts an afforestation project, is an interesting case. Here, forests are planted as a design act and as part of a larger transition. The afforestation is done in the low-lying southern Denmark meltwater floodplains, which are threatened by rising sea levels. In the case of a forest, ecologists can describe climax states, but the forest is never 'ready'. The cultural act of designing a forest slowly transgresses towards the natural forces that shape and reshape them. In these drawings, we see how the system develops a balance between decay and rejuvenation, supported by a root system that also becomes a larger whole in the drawing.

Ironically, as illustrated, holding space and time together in drawing is not difficult per se. Current practice may be hesitant to embrace time-based drawings for all sorts of reasons, but today's teaching has started to do so (de Wit et al. 2022)—and rightly so, as integrating it into the way students learn to draw future landscapes is both fundamental and feasible. As said, it starts with seriously dating drawings to say, 'this is the landscape as we expect it to be twenty years from now'. Building on that, it is only a small step to make a *series* of drawings for different moments in time. In technical terms, or seen in the evolution of representation, this is a minor step.

Current Societal Debate

Today's society faces many complex transition processes—processes that, to a large extent, are interwoven and interdependent. At the intersection of such transition processes, the short term and long term, next to the small scale and big scale, relevant landscape design issues are at stake. In my low-lying country, a small urban extension may seem a harmless project to be built in the coming years, but if positioned in a long-term future view of a delta in the context of sea-level rise, it may be a wrong choice or perhaps should be a design that can evolve with a potentially risky future. This consideration has to be made even for such a small plan, but in fact, it has to be seen in a larger frame that evaluates the future of a densely populated delta and discusses how interventions in landscape can prepare for change, particularly for change with a high level of insecurity. In these complex situations, full of uncertainties, drawings can help navigate the possible scenarios. It is in the context of today's important societal challenges and the need for high-quality

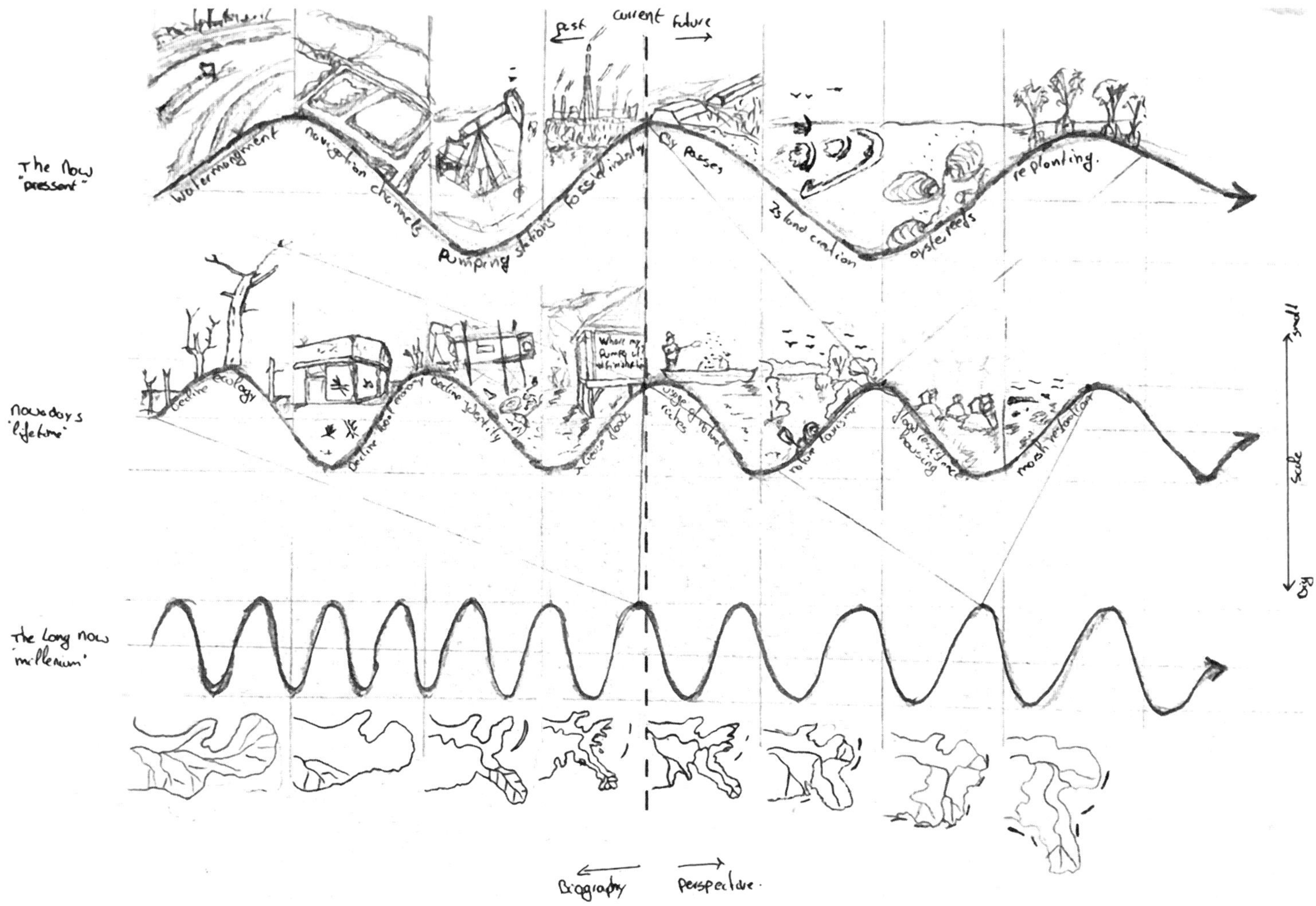

Figure 10 Workshop drawing (Marijn Vreriks, 2024). First sketch on the future of the endangered Mississippi Delta. In fact, Vreriks applied the timescales as proposed by the Long Now Foundation (see text) in this diagram. By that, he aims to position design interventions in landscape in relation to the three time dimensions.

future landscapes that I see a strong position for drawings that are capable of holding space and time together. Such drawings offer insight into the forces at work, the interdependencies, the possible scenarios, and the landscapes that may come out.

Working on his thesis on the future of the endangered Mississippi Delta, Marijn Vreriks applied the Long Time Now timescales in a diagram (Figure 10). Even if this was a very preliminary attempt, I appreciate this drawing, as it aims to position design interventions in landscape in relation to the three time dimensions. In fact, analytical observations of the processes at work historically, now, and in the future are combined with speculative observations. Reading these processes at work, how would conditions be shaped for design interventions? Subsequently, how would these interventions alter or influence these processes, and how would they possibly give rise to new interventions? Obviously, such a graph should always be read in combination with plan drawings or maps, but already in this isolated form, it promises a smart application of thinking in time in complex social and physical situations, as the Mississippi Delta is in a process of radical change.

Drawings in this context tend to become a laboratory in which potentially difficult questions are raised. But these questions may be crucial. If we were to draw the plan as a series of moments, what would be the *relevant* moments? This is exactly the core value of drawing time, as inevitably that choice must be related to the processes at work in the specific project. And each of these processes has its own timescale, rhythm, and predictability, from the 'these days' scale to 'our time' or to the 'long time'. Drawing a plan for a coastal defence project combined with parks, an urban open space, and an event area would require knowledge of the rhythm of ebb and flood, the probability of heavy storms, and sea-level change in the long run, combined with the urban rhythm weekly, seasonally, or within larger trends. This would require expertise, insight, and, more than that, managing the unknown. It may lead to a much more complicated design process, but it also works the other way around. If one starts a drawing that is inherently time based, the drawing would invite or even force the designer to explore the systems at work. Here, the distinction between the drawing as an object and drawing as a process is very relevant—long before communicating a plan or building a plan, sketch drawings in the process of ideation can help the people involved grasp what is at stake. Drawings, in that sense, also serve as meeting spaces in which comments can be given, discussions can be held, and agreement can be found. It is in this arena that space and time can find each other. Drawings in which time and space are held together can and should be made.

Image Credits

Figure 1: Lena Knappers (independent urbanist/researcher) and Bram van Ooijen (UFO urbanism), entry for EFL Climate Justice Call (2023)
Figure 2: Buro Lubbers (1998)
Figure 3: Anouk Vogel (realised 2009)
Figure 4: H+N+S landscape architects (realised 2019)
Figure 5: Violaine Forsberg Mussault (landscape architect and PhD candidate Oslo School of Architecture), PhD research (2023)
Figure 6: Louis Vanhaverbeke (choreographer/performer), Workshop Dancing Drawings, DasArts' Amsterdam School of Arts (2013)
Figure 7: GROSS.MAX, competition entry (2011)
Figure 8: MDP Michel Desvigne Paysagiste (realised 2000)
Figure 9: Johanna Bendlin, thesis, University of Copenhagen (2020)
Figure 10: Marijn Vreriks (Faculty Architecture and the Built Environment, TU Delft, 2024)

References

Blau, Eve, and Edward Kaufman, eds. 1989. *Architecture and Its Image: Four Centuries of Architectural Representation*. Montréal: Centre Canadien d'Architecture.

de Jong, Erik, Michel Lafaille, and Christian Bertram. 2008. *Landscapes of the Imagination: Designing the European Tradition of Garden and Landscape Architecture 160–2000*. Rotterdam: NAi Uitgevers.

de Wit, Saskia, Inge Bobbink, and Noël van Dooren. 2022. 'Drawing Time. Editorial'. *Spool* 9 (Landscape Metropolis 3: Drawing Time): pp. 1–4.

Corner, James. 1992. 'Representation and Landscape. Drawing and Making in the Landscape Medium'. *Word and Image* 8(3): pp. 243–275.

Corner, James. 2009. 'Agriculture. Texture and the Unfinished'. In *Intermediate Natures: The Landscapes of Michel Desvigne*. Basel: Birkhäuser Verlag.

Goffi, Federica. 2007. 'Architecture's Twinned Body: Building and Drawing'. In *From Models to Drawings*, edited by Marco Frascari, Jonathan A. Hale, and Bradley Starkey. Abingdon: Routledge, pp. 88–98.

Halprin, Lawrence. 1969. *The RSVP Cycles: Creative Processes in the Human Environment*. New York: George Braziller.

Hirsch, Alison. 2011. 'Scoring the Participatory City. Lawrence (and Anna) Halprin's Take Part Process'. *Journal of Architectural Education* 64(2): pp. 127–140.

Leatherbarrow, David. 2009. *Architecture Oriented Otherwise*. New York: Princeton Architectural Press.

Loudon, John Claudius, ed. 1988. *The Landscape Gardening and Landscape Architecture of the Late Humphrey Repton Esq.* London: Forgotten Books. (Originally published in 1840, Edinburgh: Longman & Co.).

Rosenberg, Daniel, and Anthony Grafton. 2010. *Cartographies of Time: A History of the Timeline*. New York: Princeton Architectural Press.

Schön, Donald A. 1983. *The Reflective Practitioner: How Professionals Think in Action*. New York: Basic Books.

The Long Now Foundation. n.d. *The Long Now*. Available at: https://media.longnow.org/files/2/LongNowDiag.jpg.

Tufte, Edward. 1990. *Envisioning Information*. Cheshire: Graphics Press.

van Dooren, Noël. 2010. 'Speaking about Drawing. An Exploration of Representation in Recent Landscape Architecture'. *Topos* 80(43): pp. 43–54.

van Dooren, Noël. 2017. *Drawing Time. The Representation of Change and Dynamics in Dutch Landscape Architectural Practice after 1985.* PhD dissertation, University of Amsterdam.

van Dooren, Noël, and Anders Busse Nielsen. 2018. 'The Representation of Time: Addressing a Theoretical Flaw in Landscape Architecture'. *Landscape Research* 44(8): pp. 997–1013.

Waldheim, Charles, ed. 2006. *The Landscape Urbanism Reader*. New York: Princeton Architectural Press.

Zerubavel, Eviatar. 2003. *Time Maps: Collective Memory and the Social Shape of the Past*. Chicago: The University of Chicago Press.

Mapping Fleetingness

Francesca Ceola

On Shifting Spatialities of Forcibly Displaced People in Lagos

The conceptualisations of both forced migration and the spatialities associated with it—emblematically among them, the 'refugee camp', as broadly understood—rest on an assumption of the *temporariness* of the condition of coerced displacement. This falls along the lines of 'it is not going to be forever'. For the project *Architectures of Asylum*, I study the ongoing, unfolding migration cycles of internally displaced people (IDPs) from northeastern Nigeria in and across Lagos that produce spatialities of emplacement conjugated on a spectrum of temporal un/certainty. This kind of research often focuses on the spatial qualities of forced displacement's spatialities, asking questions such as: what kinds of places do forcibly displaced people make for themselves or does the humanitarian regime provide/prescribe for them? What spatial sovereignty is afforded to the forcibly displaced? What spatial tactics are set in place to 'make do' with temporariness? The temporal values of the same spatialities are often overlooked. Aligning with this contention, as I discuss in the following pages, IDPs' emplacement in Lagos calls for spacetime attention that builds on critical analytical propositions of spacetime complexes of flight, waiting, 'stuckedness', and the everyday, to name a few (Griffiths 2014; Lems 2014; Rotter 2016; Vanyoro 2024). Thus, this chapter embeds the methodological and empirical challenge of grappling with the spacetime continuum of what I refer to as *shifting* IDPs' spatialities in Lagos within critical and creative mapping. That is, in Lagos, where IDPs who arrive in the city undergo multiple emplacement and displacement cycles within the city itself, the research attempts to map not so much their movements as the engendered geographies of IDPs' spatiotemporalities of dwelling and inhabitation in uncertainty (Boano and Astolfo 2020; Lancione and Simone 2021; Simone 2016; Simone et al. 2023).

Paying attention to the ongoing displacement (thus, the non-linearity of emplacement) in mapping these spatialities—expansively envisaged—I grapple with the possibility that the IDP settlements we[1] have mapped might disappear, as private capital and urban planning might clear people and things from areas of real estate development interest. Considering the geophysical fluctuations of the ecogeological systems in which IDP settlements are embedded adds variables to the temporariness of IDPs' shifting spatialities, and thus points to the complex and dynamic geographies of ephemerality that require pushing the limits and possibilities of mapping as methodology and approach. In tandem with the discussions brought forward by my colleague and teammate Qusay Amer in this volume, I expand on the spectrum of architectures of asylum with which our two parallel yet entangled research endeavours engage. With the notion of 'architectures of asylum', we contemplate the broad range of spatialities provided for and self-provided by forcibly displaced people (i.e., refugees, asylum seekers, IDPs, urban refugees, and forced migrants) vis-à-vis the normatively dominant 'asylum regime' that, in legal terms, frames forcibly displaced individuals' eligibility for spatial agency. Through a case study of IDPs in Lagos, I examine spacetime map making of some of these spatialities as a process attuned to time (hence, to fleetingness and ephemerality) rather than an attempt to grasp and fix time with space.

1 The mapping team was composed of a heterogeneous group of PhD researchers, local nongovernmental organisation workers, slum dweller activists, and IDP representatives. We counted on partnership and long-term collaboration with the University of Lagos Centre for Housing and Sustainable Development and the Pro-Poor Research Cluster. I use 'we' to refer to the collective empirical fieldwork, and 'I' for the speculative and analytical ideas proposed in this chapter.

It is worth taking time to consider the implication of bringing up fleetingness and ephemerality and the semantic field they mobilise in various spatial research, as well as how it informs my positioning within the study of space and movement and any attempt at 'grasping' space with movement. At the conceptual level, Cresswell's (2006) *Metaphysics of Fixity and Flow* illuminates the sedentary bias that has historically entrenched spatial research—as well as the study of mobilities, including migration, that occur in space—by coding moral geographies of place through rootedness, as opposed to dysfunctional mobilities (see also Malkki 1992). Producing and, subsequently, relying on the assumption of (fixed) spatial arrangements to study movement across them is the logical construction for the marginalisation—ontological, epistemological, and normative—of mobility (Cresswell 2006, 29–32). To think about

Figure 1 Map of self-geolocalisation *en route* to IDP settlements and around their perimeters (Francesca Ceola and İnci Sağırbaş, 2025)

migration, I find it more fruitful to follow the *metaphysics of flow* that are enriched with political situatedness and based on practice, tactics (de Certeau 1984), de- and re-territorialisations (Deleuze and Guattari 1986, 1987), and lateral movements (Simone 2020). Thus, space opens for thinking about the spatialities of temporariness, provisionality, uncertainty, volatility, and things-so-far that mobile subjects—forcibly displaced humans importantly among them—relate to. Mapping spacetime thus comes into focus as both a methodological and an epistemological approach to contend with movement. I have tackled the practical challenges emerging from this approach by creatively mapping arrangements that embrace evolving configurations, as explained in the following pages.

Displacing Lagos

This research on refuge spaces in Lagos follows the multiple cycles of emplacement and displacement of IDPs fleeing the northeast of the country who have made it to this city. Once in the megacity, IDPs navigate dense networks of socioethnic solidarities, hostilities, and spatialities of affordabilities and invisibility (Roberts and Lawanson 2023). Attuning to these dispersed geographies of sometimes provisional and other times un/secured emplacement also means understanding them within the specific urban landscape of Lagos. Here, the neoliberal expansion of the world-class and cosmopolitan megacity unfolds through 'renewal projects' that dispossess and displace the urban poor to free space for upscale development (Adeniyi Ogunyankin 2022; Ajibade 2017; Olajide and Lawanson 2022). Hence, IDPs are not alone in the circulatory movements that also affect the urban poor and slum dwellers of Lagos, maintaining them either in motion through subsequent phases of dis- and emplacement or in a state of uncertainty.

Working with this research field presents an opportunity to radically embrace the one postulate of migration studies—that things, and humans among them, move. Mapping—hence, retracing, following, analysing, and representing—moving subjects and the spatialities they carry through displacement challenges research to move along and enliven the *metaphysics of flow*, without divorcing the displacement from the political forces that produce it. In this chapter, I wish to reflect on the process of methodologically and empirically mapping a *moving complexity*—that is, the shifting spatialities of emplacements and displacements—within a volatile urban landscape (Simone 2020). To challenge the normality–crisis binary, Krause and Eriksen (2023) propose thinking through *volatility* as a flexible, often collective, way to approach uncertainties and instabilities with human/more-than-human solidarities and care (Dalal 2021). In opposition to neoliberal and depolitical understandings of 'resilience', volatile ways of life foreground people's agency in anticipating and owning fluctuations and hardships from humble and collaborative stances (Krause and Eriksen 2023, 4). IDP spatialities are thus understood as volatile landscapes of emplacement in Lagos in emerging ecologies of human and more-than-human entities in geomorphologically and sociopolitically shifting situations of hardship. While the research design hypothesises that these spatialities are marginalised in Lagos's urban economy, embracing their volatility allows for undoing their supposed marginality by recentring the research away from them.

Mapping Fleetingness

To grapple with the unity of space and time in the examination of volatility and ephemerality, the methodological approach was developed around a plurality of strands of mapping literature and practices—a continually evolving corpus of inspirations. Jorge Luis Borges's fictional short story 'On Exactitude in Science' (1999) was the initial provocation for me to think deeply about mapping, as it prickles the human ambition to create maps that are 'accurate' even as they become outdated the second they are produced. For example, the fixity of land ownership and state administrative legibility in

Figure 2 The metallic structures hanging under lorry trucks can also serve as shaded space for resting or temporary inhabitation (Pablo Quiroga Devia, 2024)

a territory—and its codification through cartography as a discipline that, for so long, has preoccupied most mapping efforts, as I discuss elsewhere (Ceola 2023)—clashes with the fluidity of landscape histories, ecologies, and economies (Aman 2022) in ways that demand that mapping be understood differently (or more expansively). Thus, my main intuition has been that, if divorced from the mapping process that originates them, maps are insufficient, as they are supposedly unifying, space- and time-fixed, and undialectical (Roberts 2016, 5). They also need to be deep and in conversation with each other in compilations that are classically considered *atlases.*

Accordingly, the work presented here draws from a constellation of mapping proposals that I continually rework through my position about politically mapping volatile landscapes and livelihoods. Among them is the hybrid mapping proposal by Baxter et al. (2021), which advocates disciplinary hybridity, embracing multiple conceptualisations of space, and working with hybrid datasets and through hybrid *modus operandi*. In addition, various strands of counter-mapping movements assert the need for epistemological openness to heterodoxies, the democratisation of the process by unmaking the mapping 'specialist' figure, and the diversification of the mapping language by allying with arts and activism (De Almeida et al. 2018; Peluso 1995; Risler and Ares 2018; Rose-Redwood et al. 2020). In the work presented here on drawing and representation, I also borrow from speculative realism approaches between creative work and aesthetic practice that broaden the spectrum of 'mapping' possibilities (Hawkins 2015; McMurray 2018; Raad n.d.; Springet 2016; Wroczynski 2011).

The following sections reflect on research conducted over two years and encompassing three fieldwork periods. I elaborate mainly on the process of experimenting with mapping the heterogeneous geographies of IDPs' shifting spatialities of emplacement. To address the challenge of capturing and rendering their ephemeral and volatile character, I use examples of two mapping instances, both analytical and field based. I thus delineate mapping as an approach to and an analytical angle for spacetime-sensitive research, casting light on the necessary attunement with environmental, urban, and community-specific spatiotemporal suggestions and transformations.

Dis/appearing Spatialities of Emplacement

The first mapping from the research I present reflects on (self-)geolocalisation, for self-orientation within the present and for replication in the future. Practically, to grasp the spatial distribution of IDPs' emplacement across the city, we visited and geolocated on digital mapping apps the IDP settlements —both clustered within themselves and scattered among other urban populations—while listening to and recording IDPs' stories of ongoing displacements since their arrival in Lagos. Using GIS technology and a survey designed in partnership with IDP and local nongovernmental organisation representatives, we collected information for the profiling of the demographic contacted —on their dispersed geographies of emplacement, and for the production of data that they can use for their own advocacy *in the future* (Ceola and Roberts 2024).

During the field visits, we produced photographs and audiovisual material of the settlements to serve as data for the secondary analysis of the materialities of the places. Capturing precise georeferences and audiovisual material to orient in the present was underpinned by the preoccupation that the places we visited would no longer be there in subsequent rounds of survey due to the threat of eviction, forced clearance, or changes in land tenure arrangements. With heightened sensitivity to the ephemerality of these places —or of any place—and the significance they carry, the audiovisual recordings captured a glimpse of their existence, while the survey data brought life to their otherwise static geolocalisation. Understanding the two components as a unified mapping process simultaneously engendered a resistance against the exclusion and invisibilisation of a large population's existence and of their spatial agency in the city (Rose-Redwood et al. 2020).

In the process, we also located and saved the bus stop where we got off, the gate through which we entered the settlements, and the spots where we explained the reasons for our visit and negotiated our presence there with the gatekeepers. Mapping the route to the settlements and our trajectories served as a reference for possibly replicating the visits in the future. This was simultaneously an exercise of recognition—understanding the city I was getting to know and the workings of an urban economy and life that were novel to me—and of mapping a field of political and socioethnic tension that upholds IDP settlements' existence and disappearance within the city—where territorial control of the urban groups among which IDPs live is asserted, materialised, and felt. With İnci Sağırbaş, we created the map in Figure 1 to elaborate on the process. Reworking the self-geolocalisation that I saved on my mobile phone, this map traces the route from the highway where we got off public transport (east-to-west road axis in the image) to two IDP settlements clusters. It implicitly marks a set of informal transportation, local negotiations, and formal introductions to the communities of IDPs and their hosts, each with their own sets of requirements and expectations. The grading of the colour of the two trajectories—from lighter on the highway to darker at the furthest point of the trajectory—highlights the processual emergence of the points I saved along the way and the perimeter of IDP settlements. Although all points are simultaneously visible on this printed interface, the shading code evokes and recalls the process and time it took behind their appearance on the base map. It calls attention to the fact that visiting settlements (quite literally, places that are home to someone) requires *time* and entails sensitivity and (capacity for) negotiation, respect for the local, and an openness to experimenting with navigating the unknown.

Volatile Inhabitations: Navigating Un/certainty in the Lagoon

The second example of mapping that copes with spacetime is more complex. It illustrates and builds on the systematisation of dwelling 'typologies' —practices of housing and inhabitation that IDPs develop—across the spectrum of geographies of un/certain settlement and emplacement. The study of dwelling invokes a broader examination of livelihoods on unstable grounds, as I present below through the collaboration with visual and plastic artist Pablo Quiroga Devia, with whom we mapped (i.e., dis- and reassembled) the spatiotemporalities of IDP emplacement in Lagos. To analyse IDP dwelling practices—which I assumed as basic spatial units for the analysis of emplacement—I elaborated on the survey results mentioned in the previous part. Upon the suggestion of the team's local collaborators, the survey embedded multiple-choice options for questions regarding settlements' composition, including the time frame of IDPs' residence there and the 'types' of housing units. Despite doubts regarding pre-determining the analysis by pre-setting the survey answers, this choice stemmed from an engagement with local knowledge holders and producers, including slum dweller activists and IDP representatives, with members of the mapping team. Indeed, empirical findings only confirmed the preset categories—self-built dwellings, rented units, and squatting situations—and the spectrum of their materialities. To unpack and substantiate these categories, with Quiroga Devia, we disarticulated, revisited, and reworked the archive of written, photographic, and audiovisual material collected through fieldwork and the survey dataset by thinking expansively and creatively about retracing meanings in these spatialities of emplacement. As I explore in depth elsewhere (Ceola 2025), through dialectical exchange and drawing (processually understood), we elaborated typologies and prototypes of IDP housing and shelter solutions. By intersecting the dwelling possibilities with the choices of construction materials and architectural practices in unstable landscapes, we thus foregrounded the temporal as much as the spatial qualities of IDPs' shifting spatialities.

Figure 3 IDP self-built housing by the ocean's coast, built of an assemblage of different materials (concrete blocks, wooden beams, corrugated zinc, plastic sheets, fabric, wood chips panels, palm leaves) (Pablo Quiroga Devia, 2024)

Figure 4 Pushing a boat through the lagoon creek where water hyacinths scatter the water surface (Pablo Quiroga Devia, 2024)

Elaborating on housing typologies and resistant or volatile materialities demands scrutiny of the land tenure in/security that IDPs, along with most of Lagos's urban dwellers, navigate and negotiate continuously within their surroundings. Here, the temporality of urban development reproduction, with a readiness to sweep away communities at any point (as has happened for Otodo Gbame, Tarkwa Bay, Oworonshoki, Mile 2, among many more), is crucial (Castro 2022; Obaitor et al. 2024; Okorji and Omirin 2019; Olajide and Lawanson 2022; Otu et al. 2023; Roberts and Okanya 2020; Sesan 2023). While being critically considerate of sociolegal and economic factors constraining IDPs' shelter-making in temporary settings, with Quiroga Devia we wished to centre the displaced individuals' spatial agency in uncertainty. Figure 2, for instance, illustrates a temporary arrangement for inhabitation that exposes one's capacity to afford or to choose their own terms of dwelling as one tactic to cope with instability. Reflecting on dwelling typologies with the artist, we analytically unbuilt the settlement and housing structures and revisited them as assemblages of suitability of needs and affordabilities of materials, as well as in terms of their volatility to geosocial and local-political processes (Baldwin et al. 2019; Clark and Yusoff 2017). Specifically, for Figure 3, we worked towards explicitly addressing the preoccupation that IDP spatialities of emplacement entangle with geophysical forces, such as coastal erosion, that determine waterfront settlements' uncertainty and underscore their shifting grounds.

One particular IDP settlement in the geophysically and ecologically specific context of a lagoon island—which was relatively 'secure' in terms of land tenure and promising in terms of social integration between IDPs and local community—was the focus of our attention and analysis. Conducting fieldwork there and retracing IDPs' dwelling practices and livelihoods required permission not only from the IDP traditional leadership but especially from the island's traditional rulers and cabinets (the island communities' leaders). It was evident that we could not work with one social group in isolation from the others without compromising local social equilibriums, as IDPs are not alone in their preoccupation with stability in Lagos. Exploring the socio-ecologies of the island residents' livelihoods, I paid attention to fruit-bearing vegetation, animal rearing, water fetching, and personal and collective mobility infrastructures, as well as the fallacies of any of these structures for residents' livelihoods as they urgently presented them to us (much due to inspiration from Kadfak and Oskarsson [2020] and Anand [2023]). Based on self-geolocalisation and the photographic and moving image archives produced on walk-alongs with different island residents, I noticed the changes in water levels at the southern and marshier coasts of the island between dry and rainy seasons and the use of the creek for mobility, farming, and harvesting, accordingly.

In Figure 4, with Quiroga Devia, we attempted to represent an instance of island residents' interdependency with the creek environment. As the physical support for their boat-based mobility, the creek is morphologically ever transforming at the pace of rainfall and sea (and river) surges. As water increases, the water hyacinth that sprinkles the creeks' surfaces proliferates, sealing the water canals with their thick bodies and thereby mobilising the islanders' physical labour to sever it to maintain the canal's navigability. While the rhythms of the rainy seasons and river and sea surges punctuate the islanders' livelihoods, issues of land use, land permits, and land rights are as pressing for the island residents' stability. The island community articulates their presence in the form of a community development project—and through collective objectives for their future. The former materialises in a map—technically, *a plan*—designed by the community with engineers and planners (which I do not have the rights to reproduce here). Generated upon multiple cycles of consultation and validation to incorporate the needs and aspirations of the community members, the map illustrates the redesign of the island settlement on the promise of land security—with the idea that displacement through dispossession will be averted. The spatial narrative construction of their community is equally a technical-political one, since it uncovers historicisations of the island communities and the formalisation of rights to the place. It foregrounds their positioning within the island's specific spatiality related to questions of indigeneity and local royalty's historical records and vis-à-vis the recent stories of clearances in other Lagotian low-income coastal communities.

Circling back to the collaboration between the artist and me as a mapping process, drawing together also made space for the spatialities of in-between—hinted at in the previous part—to emerge. In the dialogue between us, the pinpointed locations symbolically marking our trajectories to the geographies of IDPs' emplacement also generated their own geographies. For instance, during field visits, we hung out in spatialities of waiting (Gupta 2023), characterised by the refreshments purchased from street sellers or offered by local community representatives who were pondering the convenience (from their standpoint) of letting us interact with their guests—the IDPs. Examining the geographies of these *wait spaces*, we developed Figures 5 and 6, in which the defining character of these encounters' localities—the *sine qua non* factor—is the ephemeral shade afforded by architectural constructs or vegetal elements that operate as relief from climatic heat. That is, while mapping out IDP settlements' spatial distribution, shapes, land-tenure situation, ethnic history, and volatile manifestations, we also indirectly mapped the situational functionality of spatial elements that afford ephemeral shade or protection from rain, adding complexity to the intangible yet performative character of shifting spatialities.

Final Reflections

As the overall concern is the island's IDP and long-settled residents and their fleeting yet un/secured spatiotemporalities, I wish to underscore that capturing the historicised islanders' narratives came with the understanding that the projections of their own communities' futures stem from a legal assertion—that they have indigenous, customary, and factual rights to be there. This, in turn, interlaces with fear originating from the past evictions of their fellow coastal dwellers while living in *uncertain presents* (Simone 2020). In their community development project map, I read a sense of 'making do' with the temporariness and uncertainty. Somewhat closing the circle, the classically designed map of their space inverted the classical map-making canon. That is, by appropriating the language of development projects, with relevance to administrative and political actors, and the graphic skills and technical knowledge of planners and infrastructure middlemen, they produced, owned, and carefully shared the most relevant map—one that allows them to advocate for their future by navigating uncertain presents.

One main reflection was necessary for re-evaluating the utility of the mapping products generated by the research when the most relevant one already exists and is withheld by the most relevant human group. According to Wood (2015, 305), 'we think about maps as being representations of the world, but they are not. They are arguments about the world.' The many symbols and symbolic meanings conventionally translated through them only scratch the surface of the arguments conveyed by the map. Quite crucially, in the case of the island residents' maps, it projects one vision for their future with a stunning depth of detail revealed while discussing it with its makers. However, the spatiotemporal configurations between the present and the moment represented in the map are all but improbable outcomes of urban development for that part of the city. Thus, it represents a preoccupation that haunts our analytical work of coming to terms with the islanders' stories and that materially and emotionally affects their practices. In turn, it also foregrounds the ethical and normative question of who is mapping what for whom and when.

I would like to briefly underscore some reflections on the additional underlying layers of mapping fleeting spacetime. *First*, in an attempt to grapple with spacetime, the mapping of IDPs' spatialities in Lagos referred to a rather

Figure 5 Waiting under the shade of street vendors' colourful umbrellas, while the mapping team purchases refreshment drinks (Pablo Quiroga Devia, 2024)

Figure 6 Waiting under a tropical almond tree's shade (Pablo Quiroga Devia, 2024)

present and recent past narrative guided by the IDPs and their hosts' narrations. We had no intention of dismissing the impact of colonisation and the distant and recent history of Lagos and Nigeria in general on the socio-spatial processes that contribute to the formation of IDP settlements as archipelagos; thus, at times, the mapping did dive into historical records—when suggested by the island residents themselves. Importantly, our retracing of their spatialities required embracing the non-linearity of their storytelling, as (forced) displacement and emplacement are never linear (Fiddian-Qasmiyeh 2016), but rather erratic and often unforeseeable movements back and forth, forward and backward, circular and static, and all over again. Staying with the metaphysics of the flow requires deep openness.

Second, mapping as a heterogeneous process—of geolocalisation, orientation, conversation, deep listening, walking, recording, and drawing—might generate incommensurable data. For instance, posing questions to the IDP community members who have gathered in an atmosphere of expectation under the shade of a tropical almond tree canopy seemingly has little in common with geolocating the walking miles needed to get from where the IDP representatives met us to where they live. Identifying and redescribing these spatial experiences in an attempt to compose the images of IDP spatiotemporalities of emplacement gained coherence through drawing with the artist Quiroga Devia, whereby time was captured but not fixed in a drawing. The representation of a moment and the absence of representation of the moments before and after does not imply the disappearance of the latter, but rather their necessary invocation. Juxtaposing drawings as captures of spatiotemporal instances makes sense through the imagination of their proximities—spatial *and* temporal.

Third and lastly, migration as a socially relevant phenomenon produces the unfolding of space(s), and time—as non-linear—is the rhythm of this phenomenon. Thus, space *and* time need mobilising to attest to migration stories. The challenge of capturing both seems to be foreseen in quantum mechanics' postulate that you cannot know both the speed and the position of a moving particle at the same time. What we can do is approximate, as creatively and deeply as possible, fragments of an uncertain present, in which the future nests and the past is folded. Mindful of the fast-paced urban economy of real estate development, of the geophysical fluctuations to which the coastal livelihoods of the island are exposed, and of the socioethnic equilibriums tensions that can enhance or undermine IDPs' inhabitation in Lagos, approaching this through *volatile* livelihoods and shifting landscapes helped me conceptually and methodologically address the spatial and temporal fleetingness (as well as resistance) of the IDPs' and island residents' emplacement.

Funding Statement
The research was funded by the German Research Foundation (Deutsche Forschungsgemeinschaft). Projektnummer 290045248 – SFB 1265.

Image Credits
Figure 1: Francesca Ceola and İnci Sağırbaş (2025)
Figure 2, 3, 4, 5, 6: Pablo Quiroga Devia (2024)

References

Adeniyi-Ogunyankin, Grace. 2022. 'City of the Future: Lagos and the Afropolitan Imagineering Project of Owambe Urbanism'. *IJURR* Spotlight on African Futures. https://www.ijurr.org/spotlight-on/african-futures/city-of-the-future-lagos-and-the-afropolitan-imagineering-project-of-owambe-urbanism/.

Ajibade, Idowu. 2017. 'Can a Future City Enhance Urban Resilience and Sustainability? A Political Ecology Analysis of Eko Atlantic City, Nigeria'. *International Journal of Disaster Risk Reduction* 26: pp. 85–92.

Aman, Amanda, 2022. 'An Atlas of Ephemeral Geographies: Identity in the Alaskan Arctic.' Presented at *ACSA 110th Annual Meeting – EMPOWER*, 18–20 May 2022.

Anand, Nikhil. 2023. 'Anthroposea: Planning Future Ecologies in Mumbai's Wetscapes'. *Environment and Planning D: Society and Space* 41(4): pp. 683–706.

Baldwin, Andrew, Christiane Fröhlich, and Delf Rothe. 2019. 'From climate migration to anthropocene mobilities: shifting the debate'. *Mobilities* 14 (3): pp. 289–297.

Baxter, Jamie Scott, Séverine Marguin, Sophie Mélix, Martin Schinagl, Ajit Singh, and Vivien Sommer. 2021. Hybrid Mapping Methodology: A Manifesto. *SFB 1265* (working paper no. 9). https://sfb1265.de/wp-content/uploads/Baxter_Jamie_Scott_et_al_2021.pdf.

Boano, Camillo, and Giovanna Astolfo. 2020. 'Inhabitation as More-than-Dwelling. Notes for a Renewed Grammar'. *International Journal of Housing Policy* 20(4): pp. 555–577.

Borges, Jorge Luis. 1999. 'On Exactitude in Science'. In *Collected Fictions*, edited by Jorge Luis Borges. London: Penguin Books.

Castro, Brianna. 2022. 'Between a Rock and a Hard Place: Heterogeneous Governance, Claims Making and Forced Eviction in a Megacity'. *IJURR* 47(1): pp. 23–38.

Ceola, Francesca. 2023. 'Counter-Mapping and Migrant Infrastructures: Some Critical Reflections from the "Campscape" of Shatila, Beirut'. *Mashriq & Mahjar Journal of Middle East and North African Migration Studies* 10(1): pp. 137–170.

Ceola, Francesca, and Rebecca Enobong Roberts. 2024. 'Scoping Urban Informality through Co-Mapping the Data of IDP Urban Migrants in Lagos, Nigeria'. *TRIALOG* (150–151): pp. 13–22.

Ceola, Francesca. 2025. 'Exploring urban spaces of socio-environmental entanglement in Lagos: A collaboration between a researcher and a visual artist'. In *Intersectionality and the City Exploring Violence and Inequality in Urban Space*, edited by Lucie Bernroider, Anthony Miro Born, Christy Kulz, and Sung Un Gang. London: Routledge.

Clark, Nigel and Kathryn Yusoff. 2017. 'Geosocial Formations and the Anthropocene.' *Theory, Culture & Society* 34(2-3): pp. 3–23.

Cresswell, Tim. 2006. 'The Metaphysics of Fixity and Flow'. In *On the Move*, edited by Tim Cresswell, pp. 25–57. New York: Routledge.

Dalal, Ayham. 2021. 'The Refugee Camp as Site of Multiple Encounters and Realizations'. *Review of Middle East Studies* 54(2): pp. 215–233.

De Almeida, Alfredo Wagner Berno, Sheilla Borges Dourado, and Carolina Bertolini. 2018. 'A New Social Cartography: Defending Traditional Territories by Mapping in the Amazon'. In *This is Not an Atlas. A Global Collection of Counter-Cartographies*, edited by kollektiv orangotango, pp. 46–53. Bielefeld: transcript Verlag.

De Certeau, Michel. 1984. *The Practice of Everyday Life*. Berkeley: University of California Press.

Deleuze, Gilles, and Felix Guattari. 1986. *Nomadology: The War Machine*. New York: Semiotext(e).

Deleuze, Gilles, and Felix Guattari. 1987. *A Thousand Plateaus: Capitalism and Schizophrenia*. Minneapolis: University of Minnesota Press.

Fiddian-Qasmiyeh, Elena. 2016. 'Refugee–Refugee Relations in Contexts of Overlapping Displacement'. *IJURR* Spotlight on The Urban Refugee 'Crisis'. https://www.ijurr.org/spotlight-on/the-urban-refugee-crisis-reflections-on-cities-citizenship-and-the-displaced/refugee-refugee-relations-in-contexts-of-overlapping-displacement/.

Griffiths, Melanie. B. E. 2014. 'Out of Time: The Temporal Uncertainties of Refused Asylum Seekers and Immigration Detainees'. *Journal of Ethnic and Migration Studies* 40(12): pp. 1991–2009.

Gupta, Pallavi. 2023. 'Geographies of Waiting: Politics, Methods, and Praxis—A Case Study of Indian Railway Stations'. *Annals of the American Association of Geographers* 113(9): pp. 2068–2083.

Hawkins, Harriet. 2015. 'Creative Geographic Methods: Knowing, Representing, Intervening. On Composing Place and Page'. *Cultural Geographies* 22(2): pp. 247–268.

Kadfak, Alin, and Patrik Oskarsson. 2020. 'An (Urban) Political Ecology approach to Small-Scale Fisheries in the Global South'. *Geoforum* 108: pp. 237–245.

Krause, Franz, and Thomas Hylland Eriksen. 2023. 'Inhabiting Volatile Worlds'. *Social Anthropology* 31(4): pp. 1–13.

Lancione, Michele, and AbdouMaliq Simone. 2021. 'Dwelling in Liminalities, Thinking Beyond Inhabitation'. *Environment and Planning D: Society and Space* 39(6): pp. 969–975.

Lems, Annika. 2014. 'Placing Displacement: Place-Making in a World of Movement'. *Ethnos* 81(2): pp. 315–337.

Malkki, Liisa. 1992. 'National Geographic: The Rooting of Peoples and the Territorialization of National Identity among Scholars and Refugees'. *Cultural Anthropology* 7(1): pp. 24–44.

McMurray, Peter. 2018. 'Ephemeral Cartography: On Mapping Sound'. *Sound Studies* 4(2): pp. 110–142.

Obaitor, Olabisi S., Marion Stellmes, and Tobia Lakes. 2024. 'Exploring Spatio-Temporal Pattern of Gentrification Processes in Intracity Slums in the Lagos Megacity'. *Geographies* 4: pp. 231–252.

Okorji, Utchay Augustine, and Modupe Moronke Omirin. 2019. 'Risk Profile of Customary and Statutory Property Rights: Implication for Real Estate Development in Port Harcourt'. *Journal of Land and Rural Studies* 7(1): pp. 39–56.

Olajide, Oluwafemi, and Taibat Lawanson. 2022. 'Urban Paradox and the Rise of the Neoliberal City: Case Study of Lagos, Nigeria'. *Urban Studies* 59(9): pp. 1763–1781.

Otu, Mike Takim, Joseph Edet, and Emmanuel Bassey Asuquo. 2023. 'A Critical Analysis of Land Rights and Interests in Nigeria'. *Pinisi Journal of Art, Humanity and Social Studies* 3(3): pp. 257–267.

Peluso, Nancy Lee. 1995. 'Whose Woods are These? Counter-Mapping Forest Territories in Kalimantan, Indonesia.' *Antipode* 27(4): pp. 383–406.

Raad, Walid n.d. *The Atlas Group*. https://www.theatlasgroup1989.org.

Risler, Julia, and Pablo Ares. 2018. 'X-Ray of Soy Agribusiness in the Pampa and Mega-Mining in the Andes.' In *This is Not an Atlas, A Global Collection of Counter-Cartographies*, edited by kollektiv orangotango, pp. 86–91. Bielefeld: transcript Verlag.

Roberts, Les. 2016. 'Deep Mapping and Spatial Anthropology'. *Humanities* 5(5): pp. 1–7.

Roberts, Rebecca Enobong, and Ogochukwu Okanya. 2020. 'Measuring the Socio-Economic Impact of Forced Evictions and Illegal Demolition: A Comparative Study Between Displaced and Existing Informal Settlements'. *The Social Science Journal* 59(1): pp. 119–138.

Roberts, Rebecca, and Taibat Lawanson. 2023. 'Understanding IDPs in Nigerian Cities'. *Researching Internal Displacement* (working paper no. 36). https://researchinginternaldisplacement.org/wp-content/uploads/2023/05/Enobong_Nigeria-City-ID_220523.pdf.

Rose-Redwood, Reuben, Natchee Blu Barnd, Annita Hetoevėhotohke'e Lucchesi, Sharon Dias, and Wil Patrick. 2020. 'Decolonizing the Map: Recentering Indigenous Mappings'. *Cartographica* 55(3): pp. 151–162.

Rotter, Rebecca. 2016. 'Waiting in the Asylum Determination Process: Just an Empty Interlude?' *Time & Society* 25(1): pp. 80–101.

Sesan, Temilade. 2023. 'Defending Housing Rights in Lagos: A Historical Perspective on State–Citizen Contestations and Emerging Opportunities for Housing Policy Reform'. *IIED* Working Paper. https://www.iied.org/22206iied.

Simone, AbdouMaliq. 2016. 'The Uninhabitable?'. *Cultural Politics* 12(2): pp. 135–154.

Simone, AbdouMaliq. 2020. 'To Extend: Temporariness in a World of Itineraries'. *Urban Studies* 57(6): pp. 1127–1142.

Simone, AbdouMaliq, Dominique Somda, Giulia Torino, Miya Irawati, Niranjana R. Nitin Bathla, Rodrigo Castriota, Simone Vegliò, and Tanya Chandra. 2023. 'Inhabiting the Extensions'. *Dialogues in Geography* 15(1). https://doi.org/10.1177/20438206231168896.

Springett, Selina. 2016. 'Going Deeper or Flatter: Connecting Deep Mapping, Flat Ontologies and the Democratizing of Knowledge. *Humanities* 4: pp. 623–636.

Vanyoro, Kudakwashe. 2024. 'Chronopolitics: Decolonising African Migration Studies'. *Critical African Studies* 16(3): pp. 399–414.

Wood, Denis. 2015. 'Mapping Deeply'. *Humanities* 4: pp. 304–318.

Wroczynski, Emily. 2011. 'Walid Raad and the Atlas Group'. *Third Text* 25(6): pp. 763–773.

Multidimensional Cartography

Philippe Rekacewicz

Space, Time, and Imaginary

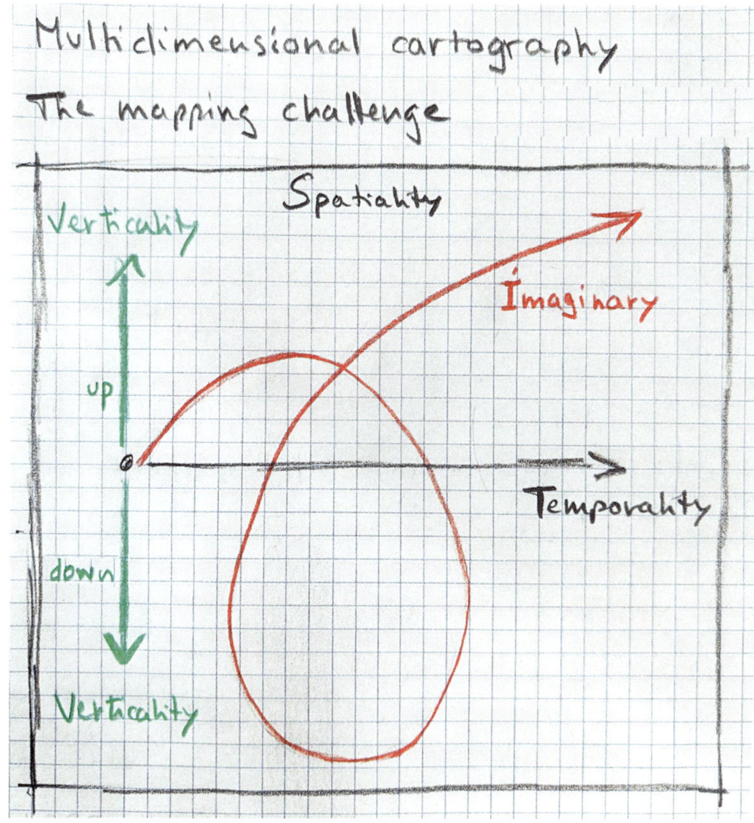

Figure 1 *La multi-dimensionalité en cartographie* (Multidimensionality in cartography) (Philippe Rekacewicz, 2024)

Prologue

This article aims to provide a response to the challenges that face cartographers, and more generally 'information designers',[1] when they are called upon to simultaneously represent or map both the spatial extent and the temporal development of data and information, whether qualitative or quantitative.

Setting out to explore a space in terms of its geography (i.e., at a single, specific moment in time) and the evolution of this geography (i.e., as a linear progression over a given period of time)—in other words, an examination of the links that bind this spatiotemporal pairing—is to enter a thought universe that is both complex and fundamental, and one which is also found in numerous other disciplines.

Swiss linguist Ferdinand de Saussure (1857–1913) defined the study of language according to two viewpoints. Synchronic linguistics 'explores the logical and psychological relationships linking co-existent terms that form a system, as they are perceived by the same collective consciousness', that is to say in a unique relationship to space. Diachronic linguistics, 'on the contrary, considers the relationships linking successive terms as perceived not by one and the same collective consciousness, but by several, which replace one another without necessarily forming a common system; it must distinguish two viewpoints, one prospective, following the course of time, the other retrospective, going back in time', i.e., in relation to evolution, to the before and the after, to the past and the present, which leads to the future. In other words, in relation to the play of time (Saussure 1916).

1 We owe this expression to self-styled *information designer* Otto Neurath (1882–1945), who in the mid nineteen-twenties developed a method of representing data in pictorial form (*International System Of TYpographic Picture Education* or ISOTYPE). He liked to say that he *literally* staged the data, transforming them into drawings, into shapes, into information.

Multidimensionalities

To these essential viewpoints, we must add two further dimensions. The first is *verticality*, for everything that swirls about the skies and space and all that is teeming below ground and deeper in the Earth's crust forms an integral part of the *world-system*. Processes not only play out on the surface, but also above and below us; airspace and terrestrial and marine subsoil are *objects*, significant geostrategic factors, in their own right. The second dimension, less concrete but no less fundamental, is *the imaginary*: those imagined worlds without which our reality would be bland and colourless. Our representations of the world, which ignore a large part of reality, involve a considerable measure of interpretation based on our imaginations, even our fantasies.

Let us consider the duality between reality and the way in which we perceive reality. Radical cartography[2] is, first and foremost, the rendering of a view—be it of a place, a region or a world—not as it is in reality but as we see it. The difference is huge: we are not mapping reality, but a human interpretation of this reality, which not only includes what we see and know, but also what our own imaginations 'tell' us. It is not a faithful representation of reality. A map, just like a novel, a short story or a play, is in fact a 'constant dialogue between the real and the imaginary' (Rekacewicz 2019).

Cartographers are therefore required to conceive their representations of the world and its complex operating processes, map the most systemic and, therefore, the most complex aspects of geography, simultaneously in these four dimensions. Of course, this *re-transcribing* of reality in drawn form produces graphical representations of great visual complexity (and frequently results in the production of so-called *spaghetti maps*, unspeakably messy drafts covered in scribblings that only make sense to those who have drawn them).

But however incomprehensible these *first sketches*, these thought pictures[3] (Marchand-Zañartu Caraës 2011), may appear to the untrained eye, they are a necessary step in the *map-making* process. The transformation of the 'cartographic intention' and the data collected for it into a *first image* represents the birth of the map, a visualisation of its structure. The next step is to articulate this visualisation using logical links and to 'dress' it in appropriate and aesthetically pleasing colours and shapes.

It is, of course, thanks to the principles of graphic semiology (Palsky 2017, Bertin 1967) and Euclidean geometry that we are able to elucidate and organise the information we seek to convey, to make it visible and accessible. It enables us to reveal significant trends and underlying processes, to provide the public with the knowledge they need to fully embrace their role as citizens, to view society, its development and shifts with a critical eye.[4]

My motivation in writing this article is to demonstrate through various examples how a well-thought-out cartographic structure combined with graphic semiology and a little geometry can help respond to the challenge of representing (a complex process in itself) the complexity of the world. This approach allows us to represent political, geopolitical and economic landscapes in a simplified (or, if you prefer, synthesised) manner without falling into the trap of over-simplification, of fooling ourselves that we can understand them at a glance. This applies equally to ordinary citizens, researchers and politicians wielding decision-making powers over their constituencies (hopefully for the best...).

Can the alternative map-making practices at the centre of our reflection, research and experimentation answer the questions posed in the scientific case made by the *Spacetimes Matter*[5] project? How can we organise multiple simultaneous spatial and temporal dimensions so that they interact in one visualisation? Can a map really convey knowledge through shape, colour and movement? How can we link a map, its title and its key to an accompanying narrative in a manner that brings all these elements into harmonious and fluid symbiosis?

In addition to being a place-finding tool, a quite legitimate *raison d'être*, or at best a simple illustration of a text, a map, as a drawn visual representation of the world, is an object that claims to make a genuine contribution to knowledge. As such, it necessarily becomes a meaningful political, social and cultural act, and that is precisely the object of radical cartography.

2 For a definition of and reflection on radical cartography and its roots, read 'Et dans les années 2000 réapparut la cartographie radicale' (How radical cartography resurfaced in the year 2000), at https://www.visionscarto.net/reappartion-cartographie-radicale, February 2012, updated December 2024.

3 In the early 2000s, art historian Nicole Marchand Zañartu collected a series of rough sketches and drafts prefiguring works by artists and intellectuals. They were the very first drawn representations of an idea, a concept or an artistic process.

4 This was the stated aim of Otto Neurath, whose minimalist visual representations were intended to popularise geopolitical and economic information, thereby making them available to the working classes, many of whom were still illiterate at the start of the nineteen-twenties.

5 The *Spacetimes Matter* project uses an empirical approach to try to understand the complex processes that preside over social and environmental change in their spatiotemporal relationships.

Sensitive, Experimental Cartography: Feeling the Space, Taking our Time Outside and Inside[6] (kollektiv orangotango 2018)

6 I owe the reflections that follow to Severin Halder, one of the members of kollektiv orangotango, an international collective of radical cartographers.

7 *The Blue Marble* is a photograph taken some 45,000 km from the Earth on 7 December 1972 by the crew of Apollo 17 as they journeyed to the Moon.

The famous *Blue Marble,*[7] the first photograph (rather than satellite image) to represent the Earth as seen from space, dates back just over half a century. This image that so fascinated humanity has shaped our perception of the planet. It also perpetuates the traditional viewpoint from which our view of geography comes, that of the spectator located outside the *theatrum mundi*, *outside* the places where things are happening. However spectacular it may be, it contributes to a process by which humanity is distanced from itself, from other living beings and from our immediate environment. The same is true of most of the maps we produce and consume, which offer us a view from without, an external interpretation of the way the world, its societies and biodiversity work. All three are indeed often *dehumanised*, disincarnated, since all trace of human life disappears as statistical formulae are transformed into abstract forms.

The idea, therefore, is to get closer to the world and to humanity, to take a closer look at the environments that surround us, the people with whom we interact—whether we love them or hate them—and to wander about the spaces that we inhabit, that we use and that we *produce*. From these explorations, these geographical experiments made from the inside, we might then embark on a *lived* cartography of these places, not from 'outside' like external observers but from 'inside' by giving voice to our feelings, our sensations and our perceptions and identifying their sensitive and emotional features? And if we find ourselves in our offices rather than in the field, we might rummage through the forgotten, personal corners of our memories to reconstruct the life stories—or simply any stories—that lie buried there. And why not represent this sensitive knowledge on a map? This is precisely the purpose of a cartographic practice that we might call *experiential*, even *sensorial*. The making of such maps reveals geographical spaces that, quite literally, form part of us and involves using all our senses and sensations: sight, hearing to pick out sounds, smell to discern fragrances, touch to feel textures and even (why not?) taste should we find ourselves wandering among vegetation.

Mapping the traces we leave in space, the paths that our lives follow—intermingled, perhaps, with our imaginary worlds—means daring to reveal the invisible, exploring the places and areas where we live from different angles, perhaps even discovering places we never go, and wondering why. It is a way of *reappropriating* geographical space, taking greater control of the environments we inhabit, since we can now see them immediately and in the round. As we explore in the field, this visual approach from the inside often serves up surprises, a new discovery or a new sensation, and this constantly renewed, iterative process invites us to seek out new methods, new graphic and semiological modes of representing and restoring our experience of using, inhabiting, exploring and producing space.

These maps show places (regions, locations, spaces) through the lens of the close, intimate relationship that exists between factual (mobile and immobile) elements and the emotional experiences that these elements provoke. *Creative experiential maps* enable us to present a wider variety of viewpoints, offering the public *rehumanised* images of the places we live in. They offer an alternative to the conventional maps from which people and feelings are generally absent. They enable us to better understand how a place operates in real and systemic terms and so, potentially, to change and reorganise them. They are a sort of in-between: both a record of an existing situation with its inherent disadvantages and a projected dream of a more pleasant, friendly, and practical space. They show how we inhabit space at a private level and so prefigure our—all too often utopian—spatial planning desires. They may offer a new way of developing and planning our towns and cities and, arguably, of reclaiming a sort of sovereignty over the public realm (reminiscent, perhaps, of the spirit and actions of the Reclaim the Streets movement[8]).

8 Reclaim the Streets (RTS) is a collective formed in the United Kingdom at the start of the nineteen-nineties comprising activists with a shared ideal of community ownership of public space.

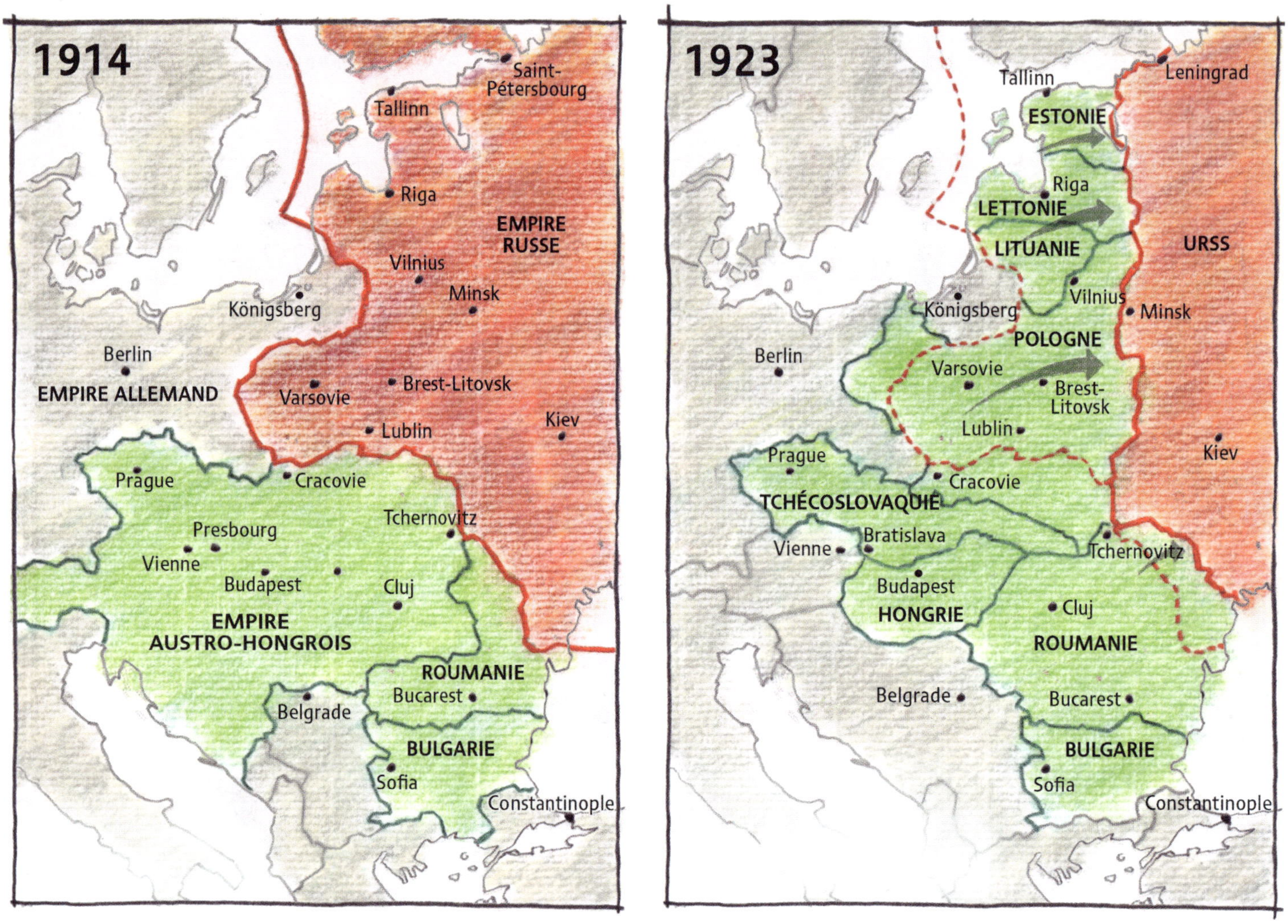

Following on from these theoretical and conceptual reflections, I would like to present the results of some of these unconventional practices. I will set out a number of potential *cartographic solutions* for transforming complex stories with their interlocking and overlapping threads through time and space into comprehensible images. These images can help us rediscover old knowledge from new viewpoints and illuminate previously unknown chapters of history and geography. This exploration, an introspective ramble through a series of maps that show how the world works, will be multi-scalar. It will start at altitude, in space, with a global vision of the Earth's continents, descend progressively to the height of our highest-flying birds, and finally encapsulate individual neighbourhoods and homes.

Making Europe Dance: The *Border Waltz*

Staying in the global dimension for a little while longer, let us look at another cartographic solution that deals fairly well with a different type of dynamic spatiotemporal complexity. This example represents a space in constant motion, a space that moves in many different directions over a long period. Trying to map the evolution of Europe's nation states through the twentieth century on a single map would only serve to add confusion to an already tangled mess of straggling lines. One solution is to identify breaks in this long narrative, as if deciding how many chapters to include in a book, and to produce a *collection* of as many maps as are necessary to provide a view of the shifting borders that shaped Europe in the twentieth century—one that is both clear and as immediate as possible. The result is a quadriptych, an

Figure 2 The border waltz (Philippe Rekacewicz, 2013)

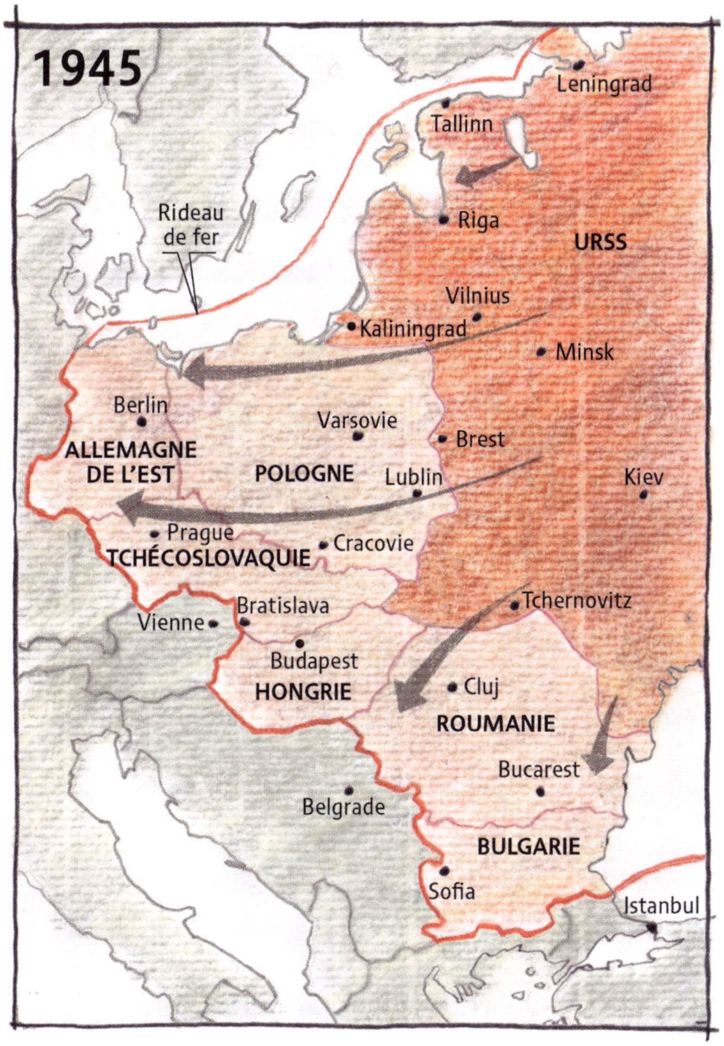

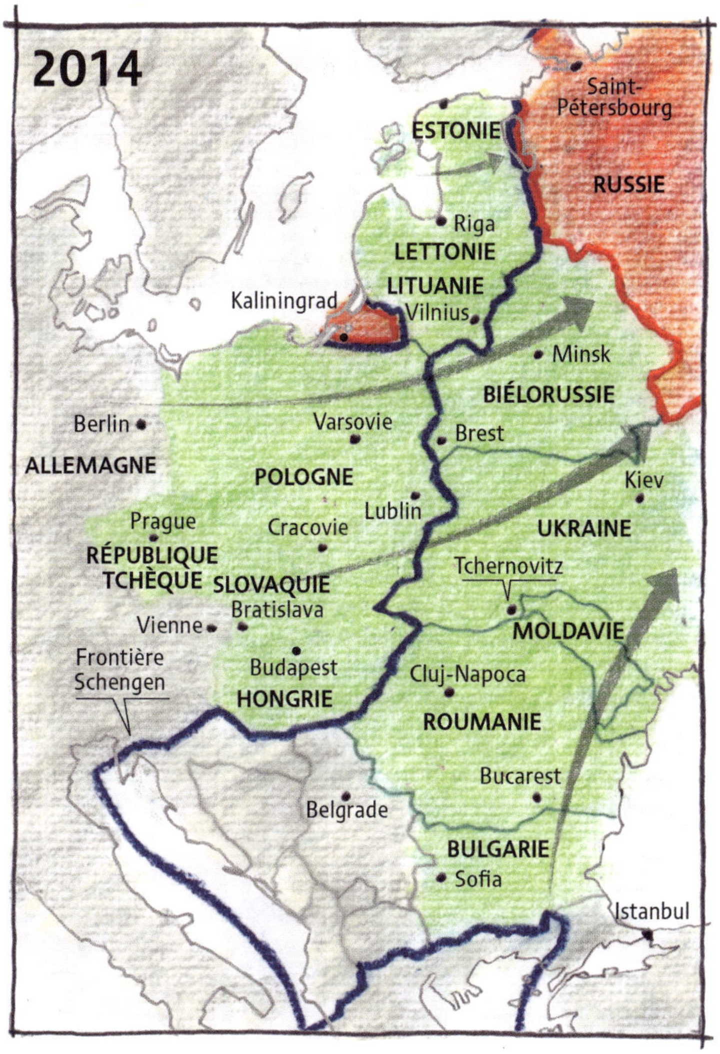

ensemble that can be considered as *one single map*. It takes the form of a visual composition comprising four intimately interconnected modules, something like a fold-out brochure[9] or leaflet. We start at the dawn of the twentieth century, just before the onset of the First World War, the event that marks the start of this frenetic dance...

1. 1914 – On the eve of the WW1, only three of the four empires that had fought for supremacy during the nineteenth century remain. The Ottoman Empire has been almost entirely dismembered, not to the benefit of the great powers (Russia and Austro-Hungary) but by the creation of new, small states.

2. 1923 – The *Treaty of Versailles*, specifically the five peace treaties concluded in 1919 and 1920, enshrines the demise of the Austro-Hungarian Empire, allowing a number of small and medium-sized nations to gain independence alongside the great powers. This is the first national fragmentation of Central Europe in the twentieth century. With the advent of the USSR, Russia—pushed eastwards in a first pendulum swing— loses the influence it previously exerted over large parts of Central Europe from Helsinki to Chișinău.

3. 1945 – At Yalta and Potsdam, in the absence of a peace treaty with Germany (the third historical empire to disappear), Europe emerges from the Second World War completely reconfigured. Its new borders, declared inviolable, will later be confirmed in the Helsinki Accords of 1975. As the pendulum swings back in the other direction, the USSR renews its attack on

9 The suffix *-ptych* in the word 'quadriptych' comes from the Greek *ptycho* meaning 'fold'.

the West, wins back and annexes a significant part of the territory lost in 1920 and turns the countries created by the dismemberment of the old empires into *satellite states*. The Iron Curtain now forms the eastern edge of Europe.

4. 1991– The USSR, the fourth and last surviving empire, implodes. The Soviet republics and socialist states of Eastern Europe regain full sovereignty, though some remain to a greater or lesser extent under the influence of Russia, which is once again pushed eastwards as the pendulum continues to swing. Russia remains in a state of *shock* until 1995, when it comes to its senses and its economic and social indicators start to progress.

Epilogue

The advantage of *breaking down* the century into a series of key periods is that we can now play with these various elements and have fun superimposing them, for example, just to see what happens. The juxtaposition of all the borderlines that existed over the course of just one century produces a completely unrecognisable political map of Europe and forces us to appreciate the degree of hyper-fragmentation that took place in Central Europe, in particular. It also clearly illustrates the singularly artificial nature of borders. A fifth map, linked to the other four, has been added to the set, creating what is ultimately a *pentaptych*.

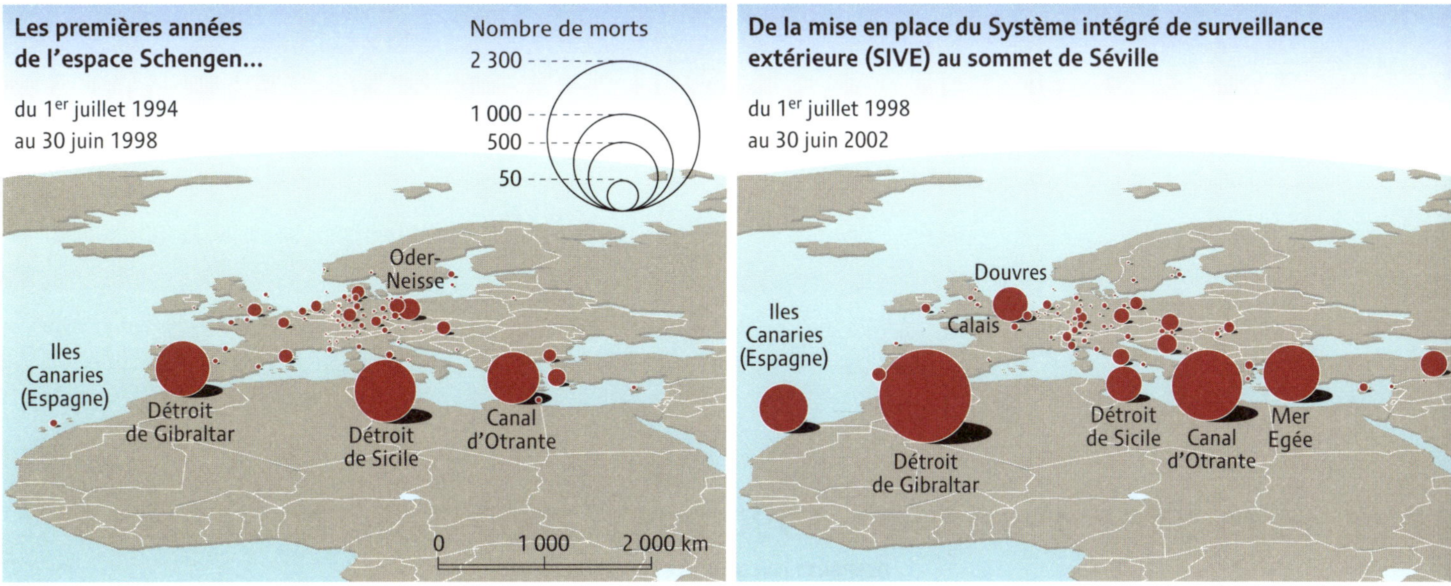

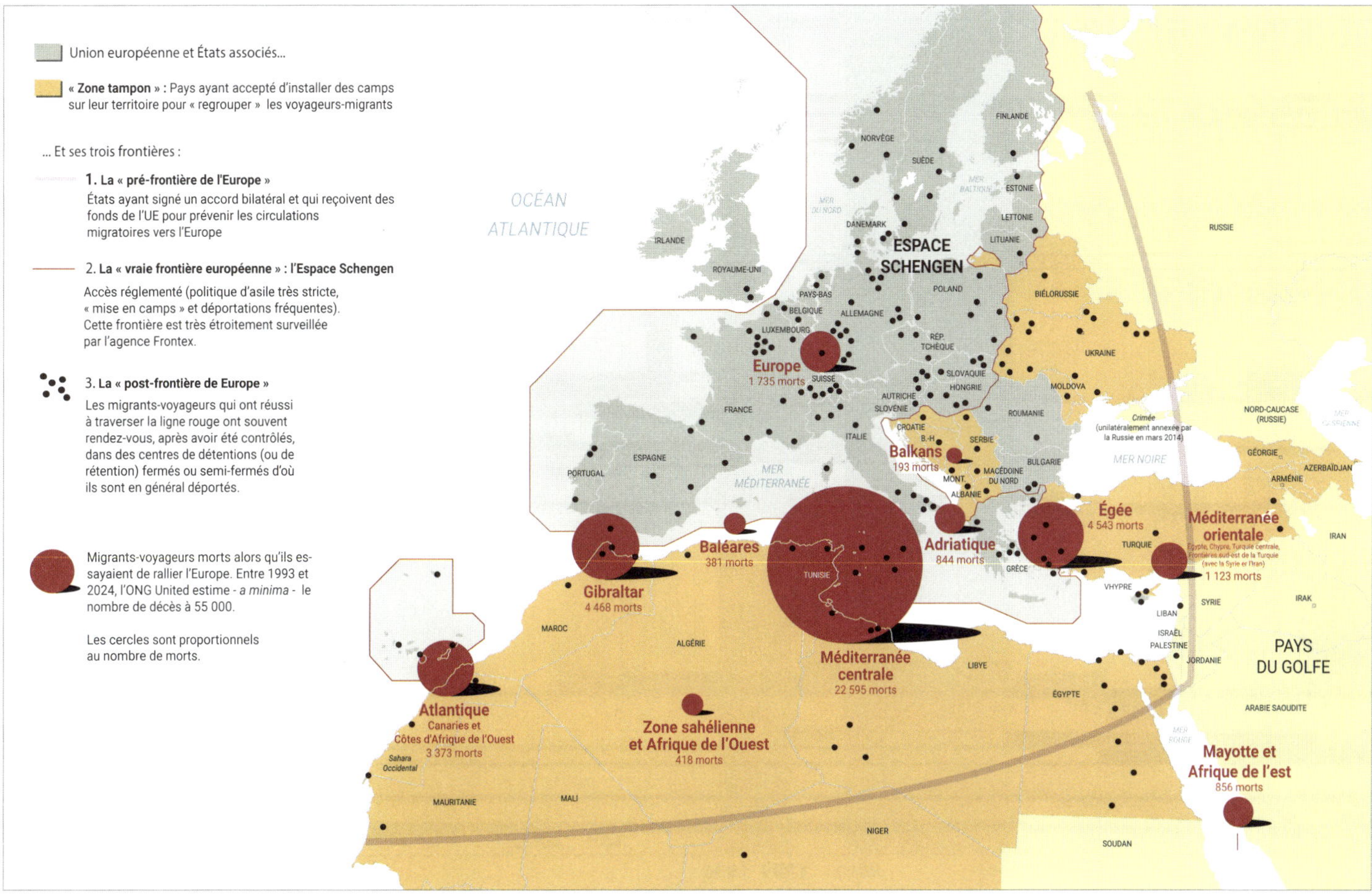

Figure 4 *Mourir aux portes de l'Europe, quadriptyque en périodes égales* (Dying at the gates of Europe, a quadriptych in equal periods) (Philippe Rekacewicz, 2010)
Figure 5 *Mourir aux portes de l'Europe* (Dying at the gates of Europe) (Philippe Rekacewicz, 2007, updated 2015 and 2021)

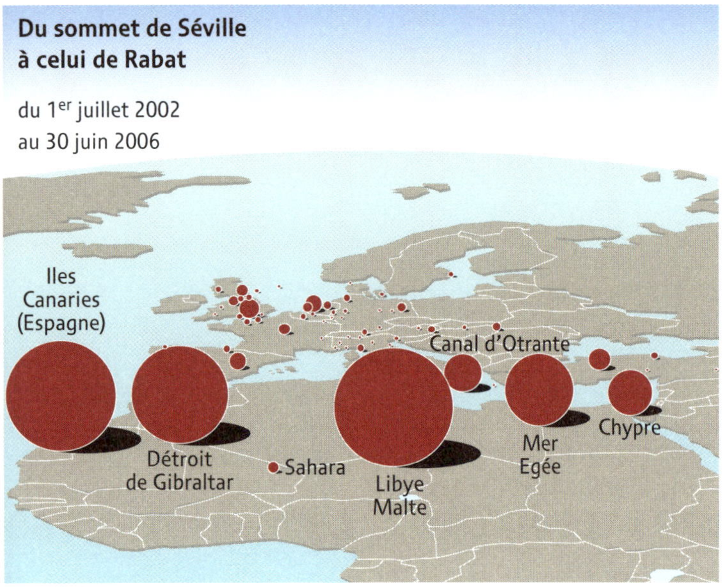

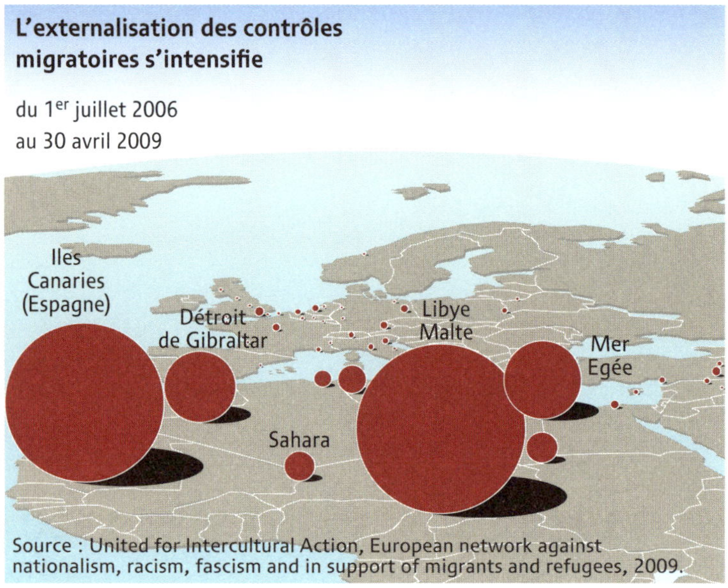

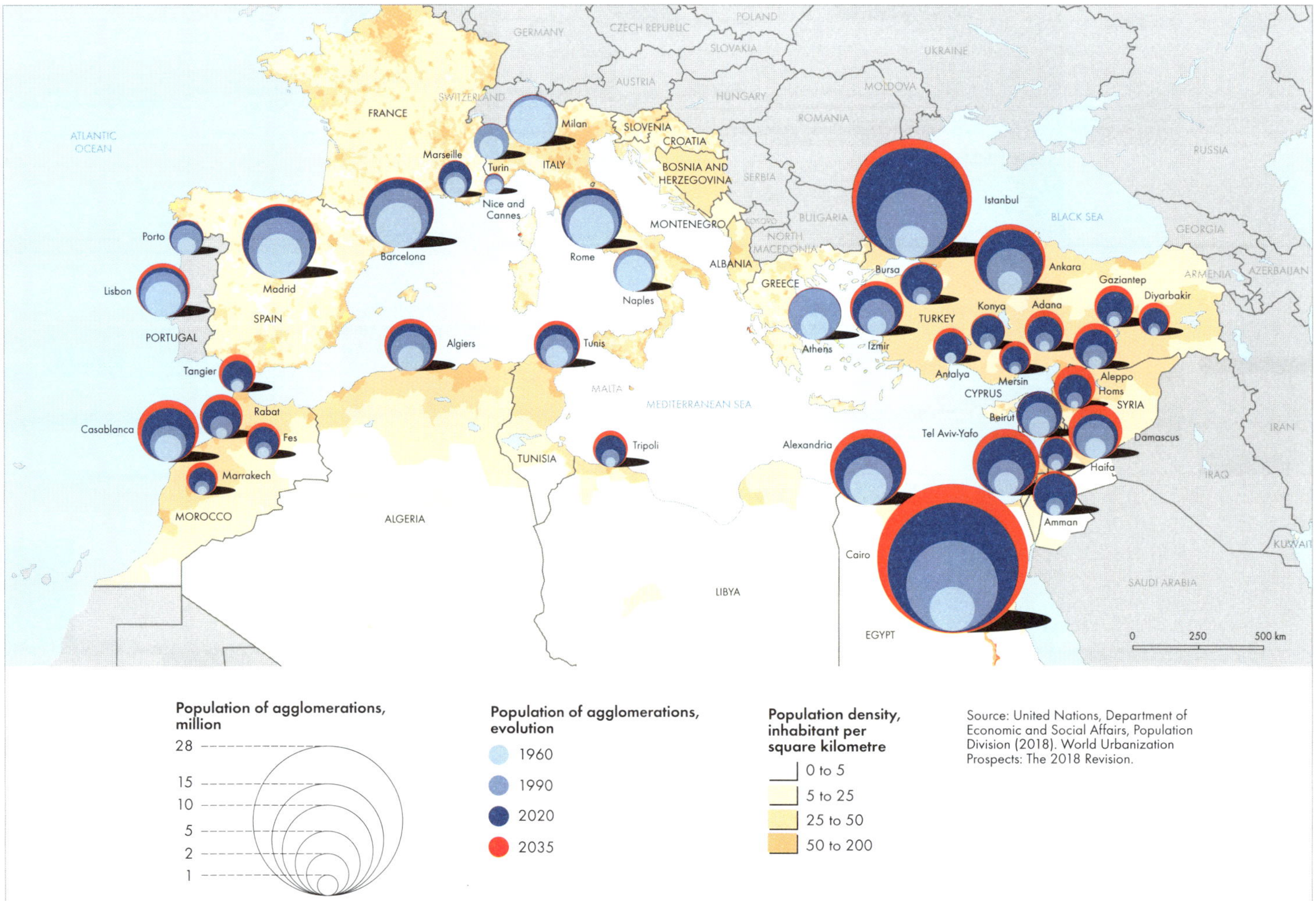

Figure 6 *Évolution des principales villes méditerranéennes*, 1960–2035
(Development of the major Mediterranean cities, 1960–2035) (Philippe Rekacewicz, 2020)

Borders that Offer No Protection

This kind of *temporal breakdown* often enables us to reveal significant trends and processes that are invisible in a simple geographical representation of a given period. The deaths on the borders of Europe map (Rekacewicz 2014)[10] provide one single, and tragic, piece of information: the locations at which some 55,000 people, at the lowest estimate, died trying to enter Europe by dangerous, alternative routes between 1992 and 2024[11] simply because Europe fails to grant them the right—a right that is, incidentally, enshrined in the Declaration of Human Rights—to travel freely to the region of their choice. This representation tells us nothing more about this butchery. To appreciate the context and the cruelly political reasons that led to this carnage, we must delve into the detail of the evolving statistics and identify three or four *chapters* (by which I mean periods of time) within this enormous database. A visualisation of deaths for equivalent periods over several decades reveals a staggering increase, showing a steadfast and inversely symmetrical correlation to the increase in the issue of ever stricter and more restrictive European directives relating to border crossings and access to European soil. These directives in turn force migration flows, even pendular flows, towards irregular lateral routes that are dangerous and frequently fatal (Figure 4, 5).

10 Read *Mourir aux portes de l'Europe* (Dying at the gates of Europe), visionscarto.net, April 2014, https://www.visionscarto.net/mourir-aux-portes-de-l-europe.

11 The ONG 'United', based in Amsterdam in the Netherlands, together with various other organisations, has been patiently collecting data on those who die attempting to enter Europe for over three decades.

But this *temporal breakdown* is not the only graphic solution we can use to map the development of events happening in the same place. It is a graphic choice made on the basis of the specific conditions that make it possible, as illustrated by this map of urban growth in the Mediterranean Basin. Though not necessarily regular, the progressions it shows develop in the same direction. There is no need to multiply the number of maps here. Simply superimposing the circles representing the four dates works very well; it provides a clear explanation of the information provided (Figure 6).

The composite cartographic installation entitled *Deux siècles de guerres, d'actes de guerre et de massacres* (Two centuries of war, acts of war, and massacres) was shown as part of FRAC Lorraine's *Zones sensibles: la carte fait-elle le monde?* (Hotspots: does the map make the world?) exhibition in Metz in 2016 (Figure 3). Shown in the same room and facing this chilling frieze, some three metres in length, which offers a metaphor for our inhumanity, a graphic allegory symbolising the hundreds of millions of war victims who have died over the last two centuries, was a series of maps intended to answer the questions raised by the image that faced them. Who supplies the arms? Who benefits from the arms trade? Who is at war today, and with whom? What effects do these wars have on the populations involved? And who, ultimately, is trying to achieve peace, and how? These illustrations are an attempt to find graphic responses to complex multi-parametric and multidimensional issues. What I was endeavouring to achieve with these maps was to *offer* the public a simplified (but not simplistic) view of this complexity, enabling them to grasp the scale of the phenomenon, how it works, what drives it and how it varies in intensity, in just a few seconds.

This frieze is a *spatial breakdown* in the form of a perpendicular structure that maps time (vertical representation) against the space in which it is unfolding (horizontal linear representation), that is to say in a dual spatial and temporal representation. Development over time (a vertical *line*, a year) runs from left to right in the various parts of the continent at war, which are *stacked* one on top of another, entirely disconnected from their geographical settings, almost as if they had been *neatly ordered* like the 'tidy cities' of plastic artist Armelle Caron.[12] The geography is merely *suggested*, taken out of context. It could almost be described as a *tempo-spatial* rather than a *spatiotemporal* representation in that the point of entry into the ensemble is time rather than geography.

12 See *Les villes rangées* (Tidy Cities), Armelle Caron, https://www.armellecaron.fr/works/les-villes-rangees/.

The angularity of the architecture facilitates a comparative, vertical and horizontal reading. Starting from the left, the eye follows a line that suddenly

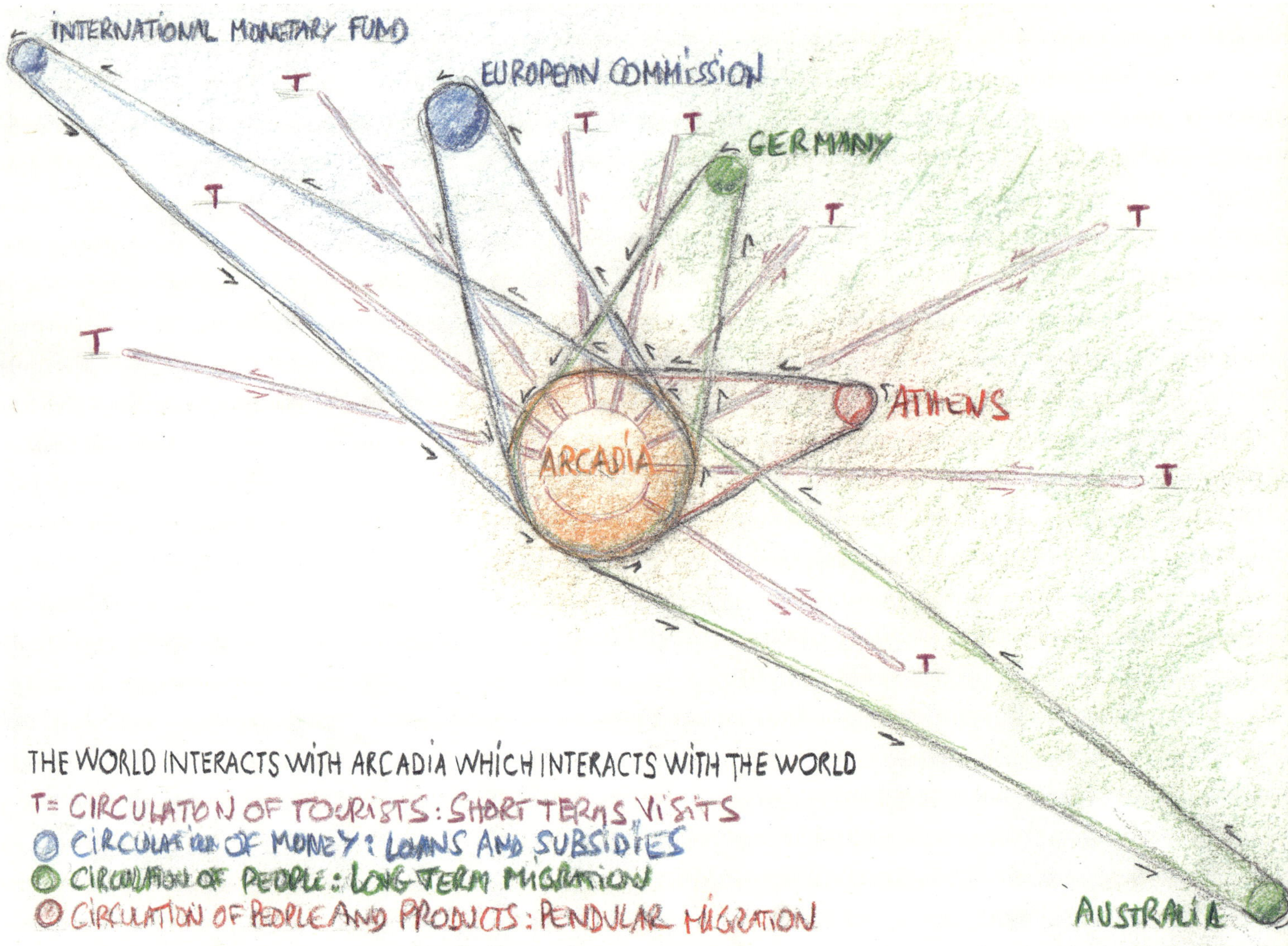

Figure 7 Arcadia as a global actor and part of a shared extended space (Philippe Rekacewicz, 2021)

becomes thicker, marking a huge explosion. The viewer stops, the eye moves up or down vertically to change location and examine events taking place elsewhere at the same time. It then continues its journey through time, either returning to the line it had initially followed or starting on a new line. Thus the exploration of two centuries of tragedy continues. Diving into this landscape, we see the global 'state of war' that is our constant reality. War is everywhere: relentless and unceasing. These wars may last anything from a few hours to several decades, claiming 60 or 60 million lives, following, succeeding and overlapping one another in one long continuum. The enduring and unchanging nature of these wars raises a question. Is it really so profitable, so economically lucrative, that nation states, other powers and, more recently, international non-state armed organisations are unwilling to renounce it in favour of other types of political management (Figure 3)?

Representing the Fabric of Urban Space

Having explored various *options* for making small-scale representations (global overviews), we now come back down to earth to take a closer, multidimensional look at the places we live in—places, incidentally, that we also explore, though often without really realising it. These may be rooms[13] or the entirety of a home, a neighbourhood, a park or a city. They are the human-scale spaces that we inhabit, use, explore but also *produce*. More than on any other scale, at this level we observe directly all that is contained or happens in them. It falls to us to draw out the most meaningful elements, the elements that best embody these places, in order to provide a meaningful picture. Here, once again, we can play around in both the temporal and the spatial dimension, helped by fertile imaginations.

13 In November 2020, working alongside Nathalie Bruyère and Imke Plinta, I co-directed a cartography workshop at the institut supérieur des arts et du design de Toulouse (isdaT), during which we spent several days developing a 'sensitive map' of the kitchen.

The cartographic corpus that I conceived and developed for the Extended Urbanisation project (Topalovic and Schmid 2023), led by Milica Topalovic and Christian Schmid at the Swiss Federal Institute of Technology (ETH) in Zürich between 2015 and 2023, enables us to illustrate these conceptual approaches.

This project was a global exploration of urbanisation processes, which extend well beyond what we perceive as urban settlements, penetrating deep into so-called rural areas but also into other spaces that we find more difficult to grasp, future development zones, for example, woodland and the sea. The team carried out fieldwork in eight regions of the world and produced detailed analyses supported by narratives, videos, photographs and critical or experimental thematic maps. They resulted in a redefinition of what is urban and urbanisation methods, changing perspective and significantly transcending the principle of the division between rural and urban space. To illustrate this approach and move away from conventional, descriptive map-making practices, the team had to use new map-building practices, in particular using *experiential* map-making methods (that originate in sensorial experiences in the field) and *conceptual* map-making methods (that allow the construction of schematic models in order to highlight complex, systemic processes simply) (Brunet 1986).

We are going to take a look at two of the places studied in this project, which resulted in the creation of original maps: the Arcadia region of the Peloponnese, in Greece, and the coastal region bordering the Gulf of Guinea between what is now Senegal and Nigeria. These maps are the result of a joint, participative endeavour by a *trio* comprising author, cartographer and the collective. These interactions, undoubtedly the most motivating and compelling aspect of the work, enabled us to compare our spatial approaches and our various artistic and graphic sensibilities (which were often different, though we always managed to reach a fair and reasonable consensus). Each of these stories is illustrated by maps at different scales, ranging from the big picture (small-scale, contextual) to fine detail (large-scale, analytical).

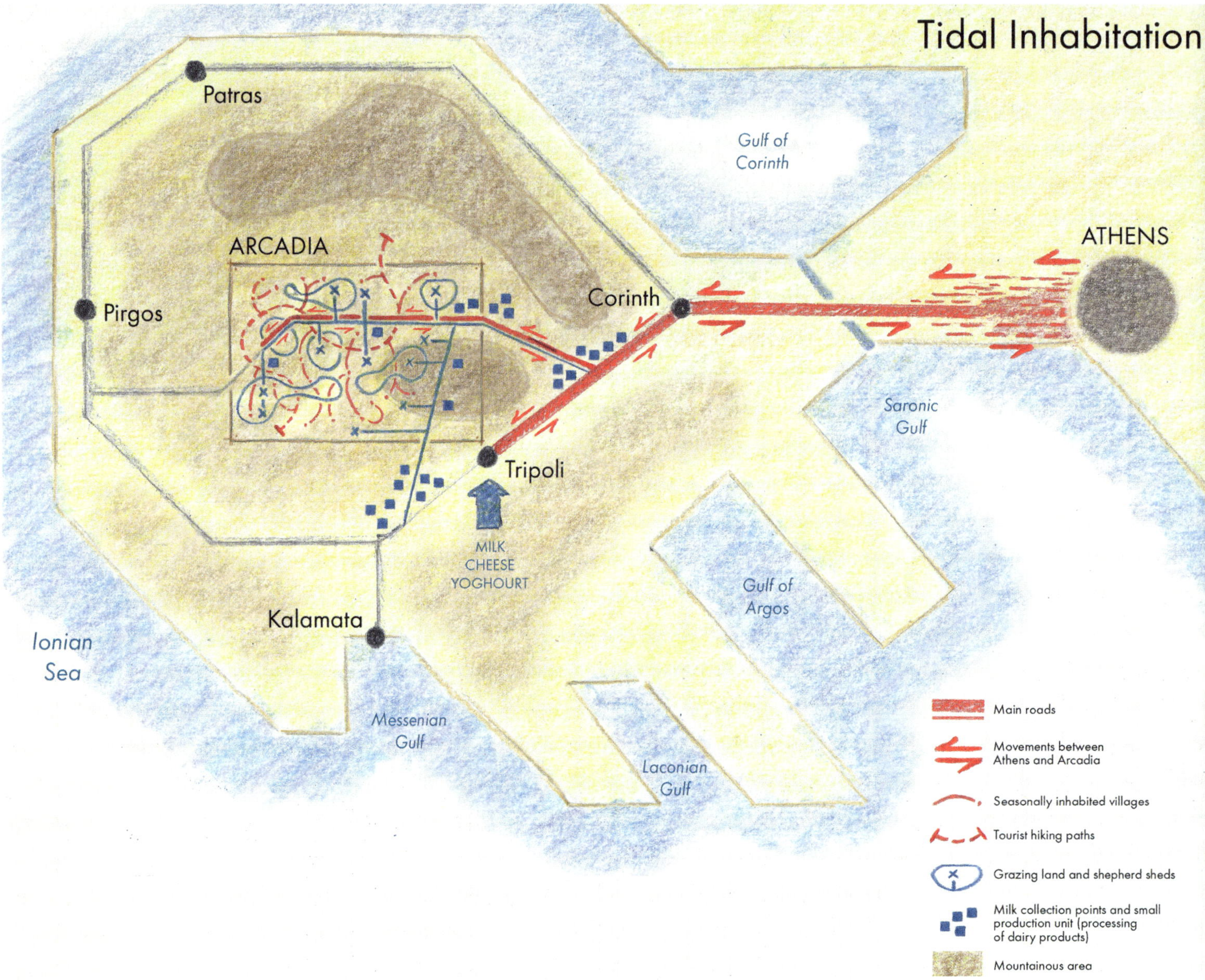

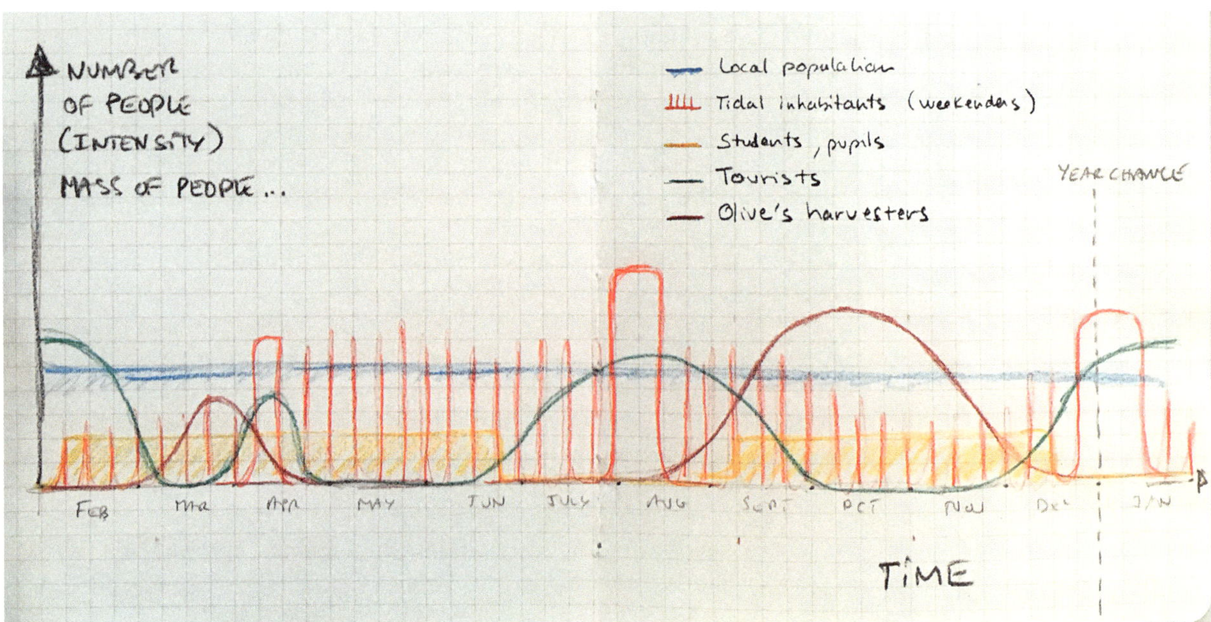

Figure 8 Arcadia, a pulsating peripheral landscape (Philippe Rekacewicz, 2022)
Figure 9 Tidal inhabitation (Philippe Rekacewicz, 2019)

Arcadia, the *Crossroads* of the World

The architect Metaxia Markaki explored the nooks and crannies of Arcadia, a small, landlocked region of the central Peloponnese in Greece, where she set out to map the dense web of ties linking its inhabitants and the places in which they live, its migrants and the region, the people of Arcadia and the outside world, extending geographically as far as Australia and, in terms of economic power, to the World Bank and the European Union. Her aim was to reveal both the systemic nature of this *landscape* and the multiplicity of links, layered one on top of the other in space and time.

The first, contextual image reproduces the dynamic relationships between the various actors in the region and the outside world. It is a map *in constant motion* with neither point of departure nor arrival, a model that dispenses with temporality or, to put it another way, a process rather than a narrative, that depicts the protagonists as they interact with the region (Figure 7).

The second image is essentially spatial. Like the first, it only addresses the issue of time through the discrete arrows that indicate the pendulum swings of displacement. The great advantage of this map is that it allows the viewer to read what is happening around the Arcadia region almost immediately. It could be described as a sort of *tableau animé* in the form of a *rough overview* (Figure 8).

Unlike its two predecessors, the third image is resolutely anchored in time. It shows the movements of the various populations found in Arcadia. The place is not stated expressly; it only appears subliminally on the map. It is simply suggested. This superposing of temporality highlights periods of intensity for each activity in relation to one another (Figure 9).

Untangling the threads of this complex story relied on a combination of semiological and geometrical principles in which nothing is left to chance: every shape, movement and direction is carefully planned and positioned to correspond to the major graphic principles of convergence, divergence, parallelism and *perpendicularism*. It is precisely this approach, this process of hierarchisation and organisation, that makes it possible to reduce *graphic noise* (i.e. reduce the perceived 'messiness' of the map), to simplify the information and, ultimately, to provide the reader with an image that is clear and lucid.

Epilogue

We are coming to the end of this exploration of experimental composite map-making, a sort of *caravanserai* that holds something of everything: representations of place and temporal evolution, a little of our imaginary worlds, shapes invented to simplify the links between elements, evocative colours. We have wandered through space and time like Broadway pantomime dancer Marta Becket (1924–2017). The founder of the Amargosa Opera House in the Nevada Desert, she wrote in gilded letters above the theatre entrance the words, 'I dedicate this theatre to the past, without which our present would be without beauty'. 'And our future without hope', she would add as she showed visitors around the theatre.

But let us return to maps. In conclusion, the next step is perhaps for us to put the human back into our maps. People are generally absent from conventional map-making practice, which revels in cold statistics that eradicate feeling and emotion. The idea is to *return* people to our maps, to *rehabilitate* them. The experimental map-making that we have discussed here is not immune from this bias and so opens up a huge potential field of research into areas that are still relatively unexplored today: finding modes of (carto)graphic representation that allow us to draw, symbolise and map the range of perceptions that people experience, the emotions they feel in their own geographical environments in the face of pleasure, beauty and ugliness, fear, joy, anger, revolt, indignation and satisfaction, even love.

Image Credits
All Figures by Philippe Rekacewicz
Figure 3: Mapping commissioned by FRAC Lorraine, 2016
Figure 6: Mapping published in *Crosslocations: a cartographic vision of the Mediterranean* by Sarah Green, Marion Lecoquierre, Philippe Rekacewicz, Rosabud (publication scheduled for 2025)
Figure 7, 8: Mapping published in *Extended Urbanisation: Tracing Planetary Struggles* by Christian Schmid and Milica Topalovic, Birkhäuser, 2023

References
Bertin, Jacques. 1967. *Sémiologie graphique. Les diagrammes. Les réseaux. Les cartes.* Paris, The Hague, Mouton: Gauthier-Villars.
Bronner, Anne-Christine, and Florence Troin. 2018. 'L'Atelier de cartographie expérimentale ou comment renouveler le processus de cartographie cartographique'. *Cartes & Géomatique* (235-236): pp. 313–319.
Brunet, Roger. 1986. 'La carte-modèle et les chorèmes'. *Mappemonde* 86(4): pp. 2–6.
Elias, Norbert. 1997. *Du temps.* Paris: Fayard.
Harvey, David. 1989. *The Condition of Postmodernity.* Oxford: Blackwell.
kollektiv orangotango, ed. 2018. *This Is Not an Atlas. A Global Collection of Counter-Cartographies.* Bielefeld: transcript Verlag.
Lefebvre, Henri. 1974. 'La production de l'espace'. *L'Homme et la société* (31–32): pp. 15–32.
Marchand-Zañartu, Nicole, and Marie-Haude Caraës. 2011. *Images de pensée.* Paris: Réunion des Musées Nationaux.
Palsky, Gilles. 2017. 'La Sémiologie graphique de Jacques Bertin a cinquante ans !'. *visionscarto.net,* https://www.visionscarto.net/la-semiologie-graphique-a-50-ans.
Rekacewicz, Philippe. 2021. 'Radical Cartography'. In *Shifts in Mapping: Maps as a Tool of Knowledge*, edited by Christine Schranz, pp. 209–232. Bielefeld: transcript Verlag.
Rekacewicz, Philippe. 2019. 'Exercice de carthographie expérimentale: dresser ma "carte d'identité"'. In *Manuel polycopié destiné aux étudiants en architecture de l'Université de Zurich (ETH).*
Rekacewicz, Philippe. 2014. 'Mourir aux portes de l'Europe', *visionscarto.net.* https://www.visionscarto.net/mourir-aux-portes-de-l-europe.
Saussure, Ferdinand de. 1916. *Cours de linguistique générale.* Paris: Payot.
Topalovic, Milica and Christian Schmid, eds. 2023. *Extended Urbanisation: Tracing Planetary Struggles.* Basel: Birkhäuser.
Wolff, Francis. 2023. *Le temps du monde.* Paris: Fayard.

Performance Drawing

Simone Rueß

Uncovering Multiple Timescapes of Nowa Huta

Introduction: Movement, Narratives, and Memories

A house façade to my right shoulder opens up via a single gate. I react intuitively as to where I move—a (re)action to the urban structure. In the archway, I slow down. The sounds from the street outside fade away—just the table, my body, the walls, and the arch in dialogue with each other. An expanse opens up in front of me, allowing me to pick up speed. I follow the directions set by the backyard streets, as if I were attracted by a magnet (Rueß, 2023b). Walking through the urban landscape, notating and observing with the sensual body is the starting point for my artistic research on space. It is an action of embodied mapping from which all my expanded drawing methods develop.

Walking is where I begin my artistic exploration and dialogue with Nowa Huta, a city of special urban and architectural-historical significance that has played an important role in the formation of social life during communist-era Poland. Constructed from scratch after the Second World War, Nowa Huta was originally designed as an *Ideal City* close to huge steelworks. With the idea 'to guarantee homogeneity' and to overcome social 'barriers, prohibitions, and divisions' (Trzeciak n.d., 2), a new city of geometric forms, with large main axes and wide courtyards, was envisioned for working-class families, who migrated there from all over Poland from the nineteen-fifties onwards. After 1990, Old Nowa Huta (Stara Nowa Huta, the oldest part of the town) underwent significant social and economic transformations, as many residents lost their jobs and general prospects. With each new generation, Stara Nowa Huta gradually gained recognition as 'a green place with good access to educational and healthcare infrastructure' and has become an attractive destination for young people and families in recent years (Kępiński n.d., 7).

To better understand the particularity of the city structure and how it is experienced by inhabitants today and in the past, I engaged with Stara Nowa Huta through several artistic interactions in the summer and autumn of 2023. I asked the following questions: how are the historical, economic, and social transformations experienced by inhabitants, and how are the changes mirrored in the context of everyday life? I conducted the performative and participative research tour, titled *Daily Routes*, by wheeling a dining table through the city and inviting passersby to share their everyday routes and memories of the city (Figure 2). These routes and memories were recorded as spoken narratives and as drawings overlaid on the printed cartographic map of Nowa Huta installed on the table. Building on my sensual walks through the city and conversations I had with participants through *Daily Routes*, I revisited six sites with my mobile table and observed daily life through performance drawings in order to understand the intrinsic logics and dynamics of places—the *Cityscapes of Nowa Huta*.

Grounded in my ongoing investigations of space as formed by daily movements, memories, and histories, these two actions—*Daily Routes* and *Cityscapes (Nowa Huta)*—helped me analyse the *Ideal City* through several methods, each standing by itself as a (serial) piece of art, but together and in relation to each other, to generate multiple perspectives on the city. In this text, I will reflect on each mapping method, discuss their specific quality and outcome, and then illuminate their interconnections and resonances with each other. First, I will talk about the sensual mapping of walking along the street with a dining table on wheels to create a performative and participatory space in the urban context, which allowed me to capture the socio-spatial dynamics of daily life in live drawings at six different locations. Residents were included in the production of knowledge by mapping their daily routes on the round dining table. The collected maps, summarised in animation, enabled me to reflect on the unfolding of different temporal continuities through 'relational drawing'. The similarities and differences of the individual narrative spaces in Nowa Huta are categorised and reflected upon through diagrams and double time graphs that show the interconnections between residents' daily routes and their individual histories. By developing individual *mental timescapes*, I was able to map together the simultaneity of place, history, and individual memories. The interconnections emerging from these various approaches were composed together in an audiovisual *city/sound/scape* of Nowa Huta—a collaboration with musician Rafał Mazur, who collected audio spheres of Nowa Huta over the last several years. The five-channel installation allows many narratives, mappings, histories, memories, and notations of the presence of Nowa Huta to circulate on the floor and four walls and in headphones, thereby creating a multisensual experience (Figure 1).

In reflecting on my artistic engagements with Nowa Huta, I begin with Kevin Lynch's (1960) proposal to approach the city's image through the researcher's observations of the site and through interviews with residents regarding their mental images of the city. I expand on Lynch's notion of the mental image by bringing my performative and participatory research into conversation with the performance drawings of other artists. 'Performance drawing', which has been established as an 'operational term' in the arts, unbounds the discipline of 'drawing' from two dimensions into a durational and spatial act. It is a processual haptic experience, either in front of an audience or documented for a future audience, and is often influenced by cross-disciplinary practices, such as dance, audio, or technology (Foá et al. 2020, 1–3). With Lydia Halcrow's *Ground Texture Recordings* and Francis Alÿs's urban interventions, I address artistic strategies of sensual walking and physically connecting with the environment through everyday practices. Understanding the motion of bodies as lines in space (Foá et al. 2020, 6), Larissa Fassler's psychogeographical maps superimpose daily dynamics, memory, and history. Ella Emanuelle's 'relational drawings' reveal the bodies' relations in motion and their relationships to the environment. Embedded in the context of the artistic exploration of space and time, this text brings together everyday life practices, physical experiences, and the connections between individual memories, collective histories, and places to form a multilayered simultaneity on the map. Similar to Christian Nold's *Emotional Mappings,* which 'display(s) individual's hidden emotional experiences in, and of, place' (Foá et al. 2020, 152), my *mental timescapes* reveal multiple, constantly changing relational images of the city, formed through individuals' remembered relations to the city's locations.

Walking and Sensually Tracing the City

Nowa Huta is a city marked by its communist history, with the years after the collapse of communism still speaking from the joints of the houses and potholes, and new cafés and restaurants telling of the arrivals of new residents and the growing consumer economy (Rueß, 2023b). Pushing a dining table on wheels through the city, from the theatre stage to the street, was initially a physical experience. I moved the round wooden table in front of me on rubber wheels over cobblestones, bumpy asphalt, and potholes; in return, its movement on the ground was 'immediately reciprocated' into my bones and marrow (***idem***). While Lynch (1960, 15) proposes extensive exploratory tours through the city to arrive at a 'subjective judgement', my body sensually absorbed the urban structure of Nowa Huta with the help of a table and its rolling, sometimes wobbling contact with the ground. In his performance *Paradox of Praxis 1. Sometimes Doing Something Leads to Nothing* (1997), Francis Alÿs uses a melting ice block to draw a disappearing line of water through the urban landscape of Mexico City, until the ice cube dissolves into the city space. Different to Lynch, Alÿs and I apply our walking bodies with objects and connect them to the ground, the environment, and its daily matters to form a temporary 'performative space' constituting a relational structure with the urban environment's life (Grünewald 2017, 10). In other words, our wandering movements—our performance drawings—constitute a temporary relation that affects not only us, the artist-researchers, but also city dwellers and other beings. My motion draws an invisible line on the urban landscape, which, through conversations and drawings, forms connections with residents' everyday interactions with the city. What begins as a solitary

Figure 1 *city/sound/scape*, audio-visual installation, five channels, headphones, 51′27″, loop, Teatr Łaźnia Nowa, Nowa Huta (Simone Rueß and Rafał Mazur, 2023, photo by Dominik Stanisławski, Dom Utopii)
Figure 2 *Daily Routes (Nowa Huta)*, performative and participatory research action, collaboration with forty-seven participants during six days in Nowa Huta (Simone Rueß, 2023, photo by Dominik Stanisławski, Dom Utopii)

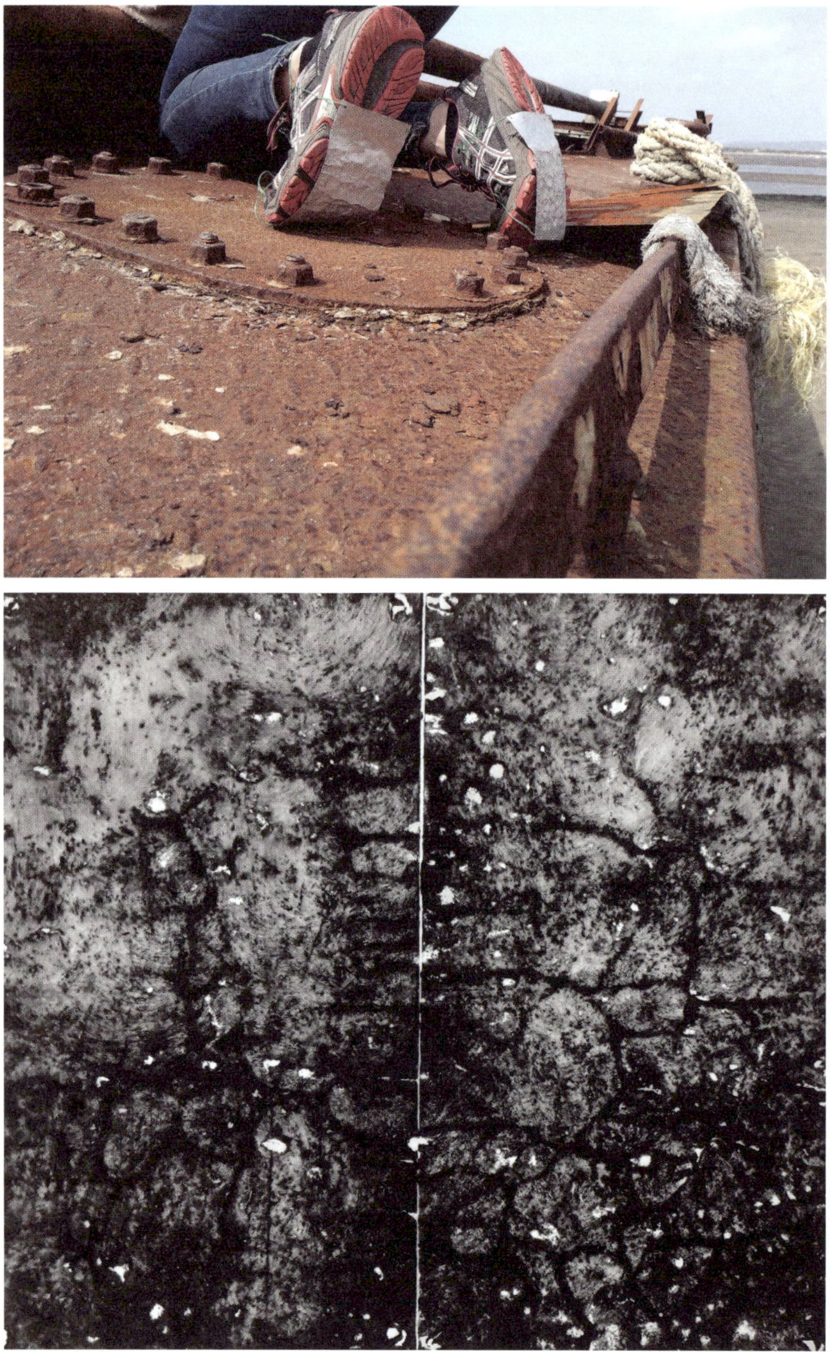

Figure 3 *Ground Texture Recordings,* 2019, Above: work in progress, photo by Al Halcrow, below: part of an ongoing series, experimental Drypoint Etching, Bideford Black on Somerset, 18 × 15 cm (Lydia Halcrow, 2019)

Figure 4 *Osieldle Stalowe* (Cityspace Nowa Huta), 20 September 2023, Performance Drawing. Above: Video edited by Hara Shin and Simone Rueß, below: photo by Małgorzata Szymczek-Karnasiewicz (Simone Rueß, 2023)

performance of walking evolves into a participatory space through conversations with the people who inhabit and move through the city.

Cityscape Nowa Huta and *Movement Space*

Based on the insights I received during the conversations and narrative mappings of residents along my first tour through the city, I revisited and approached select locations in greater detail (Lynch 1960, 15). I stood with my mobile table at six locations for a longer duration of time (approximately two to three hours), scanning the visual field and documenting the multiplicity of moments as layers of strokes and lines on a single sheet of paper (Figure 4). The live drawings were simultaneously filmed from above and later digitally sped up in videos and superimposed. Inverted to white on black, like an X-ray, the time-based images illuminate 'a view of bodies seen through and beyond their skins'—that is, a view of environmental structures seen through their surfaces (Rueß 2023a, 89). Built on my artistic practice, I used the method of overlaying drawn movements into simultaneity to reveal the so-called *movement space* in relation to the architecture, environments, and their histories. Halcrow (2022, 25) connects with and 'tunes into places' such as the Taw Estuary by *Walking* and traces the texture of the site by capturing imprints on metal plates attached to the soles of her boots. While her embodied mappings uncover her own movement by measuring 'the meeting points between her feet/body and the estuary/earth/textures through gravity and motion' (Halcrow 2022, 32) (Figure 3), my filigree drawings display the sensually scanned movements of the visual field. Our differently produced, translucent landscapes of movements reveal different experienced structures inherent to places.

The live drawings in Nowa Huta let me sensually approach the specific local dynamics in courtyards and main axes and at the boundaries between natural and built landscapes. For instance, the live drawings allowed me to understand the gender dynamics of girls and women spending more time in safe courtyards. Everyday experiences in the city are gender-specific and, even today, are associated with fears for women (Kern 2022, 16, 157). Taking up space in the city as a woman is still unusual. It is even more unusual that I, a woman who does not even speak Polish correctly and has no Polish roots, can wander surrealistically through the streets of a city marked by its transformations. In a similar fashion, Larissa Fassler uses her body and her sensual apparatus to critically scan the urban places that often conform to masculine notions of space and city. 'Her personal measurement systems and her notational tracings of experiential and archival information' confront the architectural and infrastructural environment with spatiotemporal notions in multilayered drawings (Furján 2009).

Participatory Mapping of Everyday Life

While my performance drawings, like Fassler's psychogeographic mappings, traced the current characteristics of certain locations in the urban landscape, the interviews with residents of Nowa Huta provided insights into individual stories from the present and past. On their way to carry out their daily duties and habits, passersby encountered my performative research path through the city. As an original dining table, the furniture constituted a link to everyday domestic life. Over six days, I engaged in conversation with forty-seven people, with participants ranging from senior citizens, middle-aged people, and young residents to children. I met first- and second-generation migrants, as well as new residents. A map of the oldest part of Nowa Huta lay on the round table and connected locals with their individual city knowledge, making participation more accessible.

The map of the city revealed how its built environment, the construction of which started in 1948, was meant to embody notions of the *Ideal City* built on communist principles. A pattern of black rectangles extended from the central square in the middle of the map to one side of the table and represented the geometric layout of buildings along streets and polygonal backyards at

Figure 5 *Map 10 ID21, 22 (Nowa Huta).* Above: Map, 100 × 100 cm, marker on transparent paper with black map underneath. Below: Still image, narrative mapping on a mobile table (Simone Rueß, 2023)

their backs. A few lines marked the Vistula meadows, called Łąki Nowohuckie, located right next to the central square of the Old Town (Rueß, 2023b). As the massive steelworks plant, which is now largely disused, lies outside of the chosen detail, the map predetermined a focus on the housing and living situation of Nowa Huta today and in the past. Using this map as a visual cue, the residents I met reflected on their daily spatial routines in relation to the ideally constructed Stara Nowa Huta (Figure 5), while drawing their everyday movements on tracing paper laid over the map. The table was round, so we could converse and circle around it and change perspectives dynamically. An action camera and a microphone were installed above the table to record the mapping processes from a bird's-eye view. The recorded routes, animated as emerging lines and superimposed, reveal a common, constantly moving form that summarises the continuous cycles of the everyday life of many actors—a collectively produced spatial knowledge that reveals both the logic of the *Ideal City*, on which the urban structure is based, and how, through everyday practice, this logic is reshaped.

Relational Drawing

In the animated versions I made of the participatory maps described above, lines in motion appear in relation to other growing lines (Figure 8). Built on previous works, such as *Patelnia (Warsaw, 2011–12)* (Figure 6, Kusiak 2012, 51–60), I use the figure of animation to visualise the ever-changing arrangements of the socio-spatial relations (Löw 2016, 159) in places. My animations bear resonance with *dAnCing LiNes* (Figure 7), a project by Ella Emanuele in collaboration with choreographer Simon Birch and data visualisation expert David Hunter. Emanuele's project visualises patterns and formations of moving dancers' bodies by using data-capture technologies (Emanuele et al. 2024, 615). Mechanically produced simulations of scores, superimposed on the documentation of the performances, create coloured, moving lines between bodies in video projections and reveal the rules of the underlying choreography at each location. While these 'system-based relational drawings' are connected to the spatial environmental conditions of the original live events, they uncover 'sightlines and perspectives that cannot be seen simultaneously during the live events' (Emanuele et al. 2024, 614).

Both Emanuele's and my relational drawings correspond with Massey's (2005, 95) idea of space as a sphere of multiplicity of social relations. The environmental setups, the arrangements of buildings and objects, appear in the frame of the videos as fixed—unmoving and unchanging. These environmental elements change their spatial structures over historical time. Certain places, streets, and special features of the Nowa Huta map are highlighted through the multiplicity of animated lines. For example, many daily movements by residents circle the significant Vistula meadow, which still reminds the residents of the historical rural site on which Nowa Huta was constructed after the Second World War from 1948 onward. Through the relational drawings, I found that the buildings and layout of places embody changes in history, while the cycles of daily interactions appear as a 'steady flow of becoming and change' (Gablowski 2013, 23). Under the endlessly changing character of daily activities at a place, continuity can be seen, revealed by the infinitely repeated routines and movements. There is a tension between the temporality of history and that of everyday movements that is expressed by the different rhythms of different temporalities. Bringing together the spatial layout and daily movements across space expands Lynch's *Image of the City* to an understanding that this image has multiple layers of history and that these layers are continuously changing. There is one more layer of temporality, which remains invisible in these animations but which I explored in this project, explained in the next section.

Figure 6 *Patelnia (Warsaw)*, still image from stop motion 3′07″ (Simone Rueß, 2011/2012, music by HITODAMA, Corvo Records, 2010, 3:07)
Figure 7 *dAnCing LiNes*, Above: *The Moor*, Below: *The Pier* (Ella Emanuele, 2020/22, in collaboration with choreographer Simon Birch and data visualisation experts David Hunter, Zach Duer. With thanks to the Arts Council England)

Different Types of Movement Spaces in the City

Multiple individuals move and interact on the basis of their own biographical stories. My performative setup in Nowa Huta allowed me to collect personal references to places, people, and the history of the city along with the mapping process. The participants started with a certain location on the map—with their current home, their home from the past, their favourite place, and so on—and drew a growing line that had a destination, or they made a loop and came back to the starting point. A second route, and then a third, and so on, expanded the individual maps. My interviewees explained and specified the buildings, places, and dynamics among the main streets, backyards, boundaries, and marks. Thus, I was able to gain an idea of 'their own images of their physical environment' (Lynch 1960, 15). Along their drawn lines, they reconnected to memories of the past and history, including changes and challenges. They reconstructed daily routes from their youth, from the past, or from the present (Rueß, 2023b).

I have been collecting *Space Biographies* since 2016, and I have developed icons for certain themes that allow me to structure biographical narratives. With my visual analyses I relate to Gunter Weidenhaus's *Soziale Raumzeit* who emphasises how living spaces such as the home, the workplace or vacation site 'spatialise social relations' (2015, 152). For example, interviewees would refer to their childhood or youth and construct their spatial experiences chronologically throughout their biography, while others would relate their perceptions to their body and other beings. Categories such as 'childhood-related', 'body-related', or 'nature-related' helped me specify the map types in Nowa Huta (Figure 9). According to the topics on which they focus, thematic groups of maps show similarities in their formations. People who unfold a body-related narrative, who walk with their dog or with their relatives, create daily routes that span the entire map of Old Nowa Huta, regardless of age. Participants who remember places from their childhood tend to draw dots and loose patches on the map.

Memories in Double Time Graphs

The qualities of certain habits and routines are represented in the simple shapes of lines. However, the memorised individual historical or biographical specifications along the narrative mappings had to be reconfigured with the help of double time graphs, a method that I developed between 2019 and 2022 in *Space/Biography (Fragile Memory)* to record how memories themselves change, as in the case of seniors suffering from dementia. Superimposed graphs of up to sixteen interviews of each respondent, animated along a vertical historical timeline and a horizontal narrative timeline, show how certain experiences, such as geopolitical caesuras, gain more frequent resonance. The condensation and fanning out of the animated lines between the limits of life led me to the creation of *Fragile Remembering,* a kinetic landscape with slowly moving paper stripes that materialised remembered interpersonal relationships from present and past. The abstraction from double time graphs to a landscape of memories translated temporal snapshots into space, creating an imagination of time in spatial form, which was, in fact, an artificial reconstruction of time (Bergson 1993, 25–27). The transformation of narratives to graphs and, further, to three-dimensional scapes was essential as a method for my analyses of the Nowa Huta maps.

From Graphs to Timescapes

In the double time diagrams, along the narratives of *Daily Routes (Nowa Huta)* (Figure 10), I differentiated the historical and biographical information using different colours. Yellow markers show biographical notes, such as the route to school in childhood, place of birth, and so on. However, two colours are used to distinguish historical references. Purple markers represent the historical period before 1948, when the city did not exist—for example, when participants talked about the rural area before the city was built. Everything that

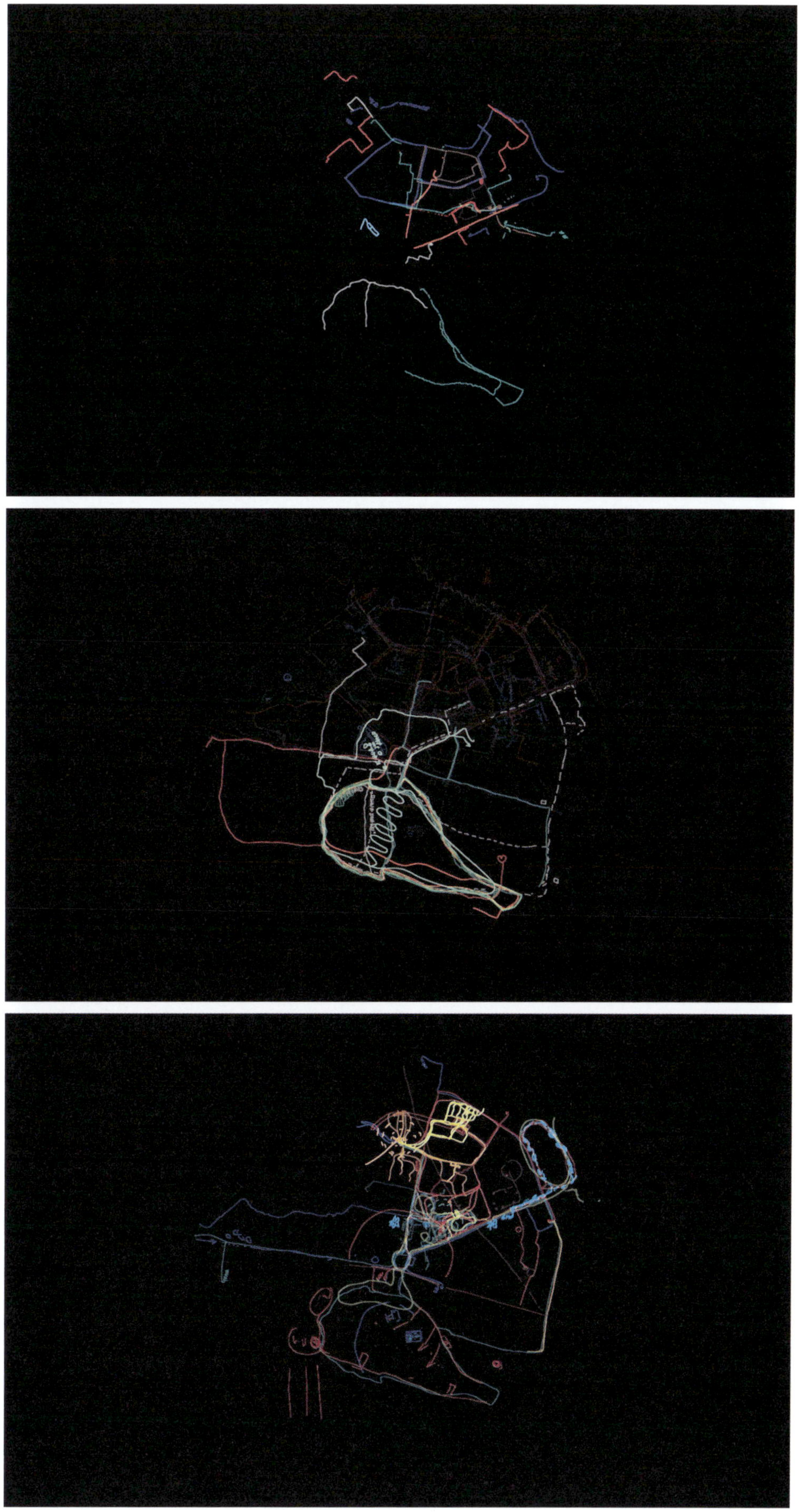

Figure 8 *Movement Space (Nowa Huta)*, still images from animation based on narrative mappings with forty-seven interviewees, part of *city/sound/scape* (Simone Rueß, 2023, edited by Hanna Grodner and Simone Rueß, sound by Rafał Mazur)

happened after 1948, when the Old Town of Nowa Huta was built, and afterwards, is marked in red. This includes explanations of the function of buildings, their architectural quality, and so on. The frequent appearance of historical markers in biographical narratives shows which time periods determine the greatest spatial-sociological impact from the interviewees' perspectives. For example, ID17 and ID04 frequently referred to the time of the construction of the *Ideal City*, but less often to the transformation of 1990, while the biographical time of the daily commute was constantly present. In the diagrams of seniors, who reported routes from the past, red historical and yellow biographical references often overlap.

My interviewees described certain monuments and talked about changes in the courtyards, now closed shops and restaurants, and newly opened cafés. Illustrating the different topics along the narratives using icons helped to identify the occurrence of certain categories (Figure 11). This bears resonance with Lynch's visualisations of certain characteristics of the city. However, marks, shapes, forms, boundaries, and locations not only explain a certain quality but are receptacles of remembered information and emotions. Therefore, I created a third coordinate representing the map of Nowa Huta, on which I localised the remembered references along the drawn route.

With a view through different layers and times, multiple historical or biographical references were drawn three-dimensionally below the daily route at the locations mentioned in the interviews (Figures 12). The further back in time the event was, the deeper the layer on the map. Besides the red and purple historical events, the green marks refer to nature, parks, and green areas; blue represents architectural references of the *Ideal City* layout; and pink refers to recent changes. With the grey in the back, each personal story is conjured as an individual three-dimensional space along Nowa Huta's geography, urban landscape, and history (Figure 12). The icons along the lines highlight the explained relationships to other bodies, nature, architecture, history, or biography, which I had previously analysed along the individual mapped lines. The spoken and drawn narratives become visually connected and related to each other.

Similar to my *mental timescapes*, Christian Nold explores in his *Emotional Mappings* residents' 'relationship with their local environment' (Foá et al. 2020, 152). He tracked the participants' journeys through GPS technology and recorded their emotional experiences along the routes through galvanic skin response-technology applied to their index fingers. The individual's hidden emotional experiences are expressed as graphs and superimposed onto the street maps (***idem***, 153). Locations are layered with individual emotions in three-dimensional shapes along the map. While Nold visualised these landscapes of emotions by employing technologies, I registered detailed historical and biographical relationships to the city through narratives, analysed first in double time diagrams to locate them later on the map in *timescapes*. While the double time graph (Figure 10) shows the frequency of time references, the *timescape* reveals the spatial spread of memories across the city map. For instance, the *timescape* of ID04 (Figure 12) shows how the purple references are spread over large parts of the map, while the red markers are placed more locally. In contrast, in the double time diagram of the same interviewee, red marks appear very frequently, and purple appears only rarely. The two-dimensional diagrams do not clarify that the references to the time after 1948 appear at the same places, while the purple refers to a larger urban area. Animating these landscapes of memories in sequences provides a differenti ated view of the chronology and frequency of remembered historical references and their spread across the urban landscape. In motion, the *mental timescapes* unfold individual topographies of the city in which movement, place, narrative, biography, and history are interwoven.

Earlier in this article, I discussed 'relational drawings' and the temporality of daily activities in relation to the duration of environmental changes. Compared to that, the *mental timescapes* relate daily activities not only to the

ID11

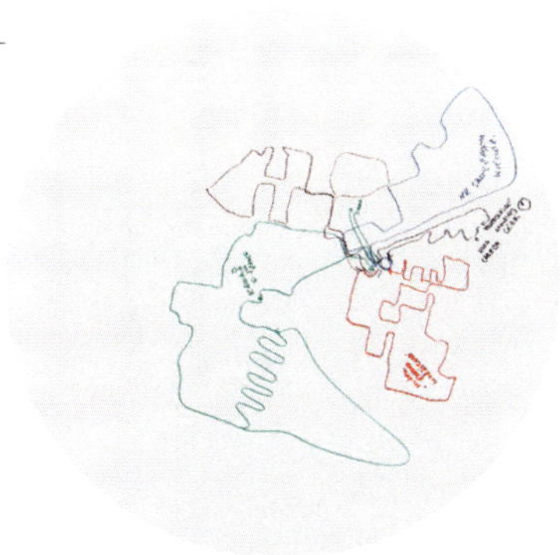

migrating to Nowa Huta in 2006

presence | body-related | nature

ID12

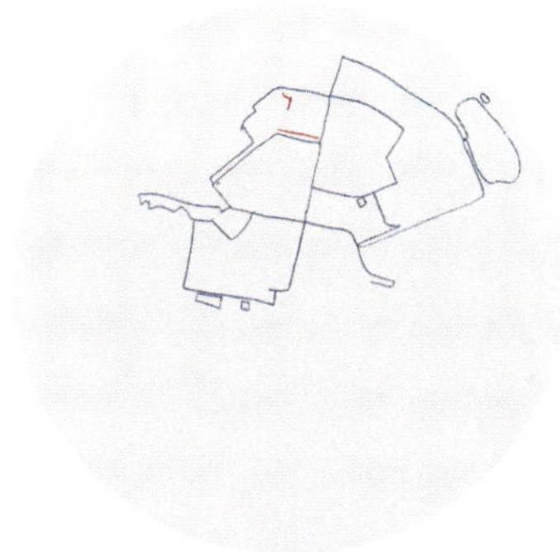

migrating to Nowa Huta in 2006

presence | body-related | objects

ID17

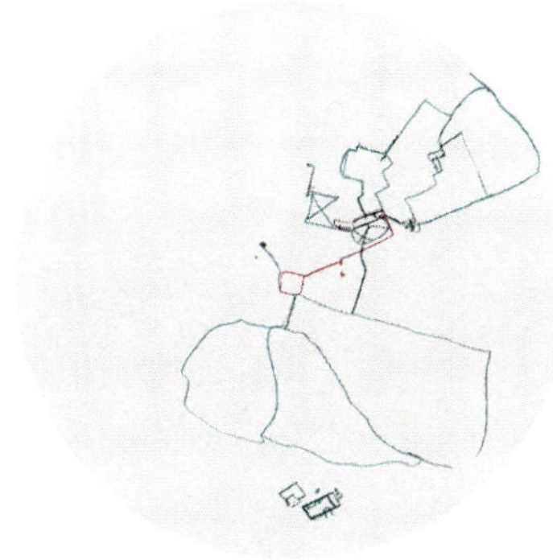

migrating to Nowa Huta in 2020

presence | body-related | NH layout

ID18

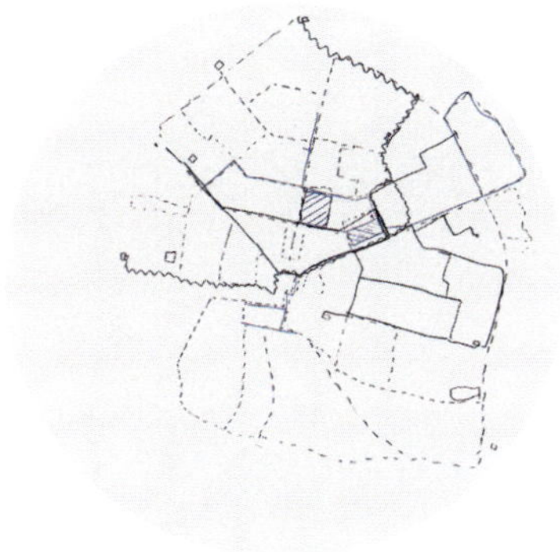

migrating to Nowa Huta
in the recent years

presence | body-related | nature

ID23

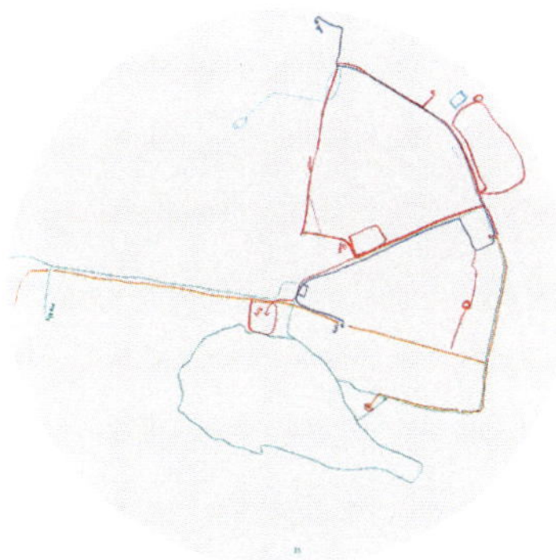

migrating to Nowa Huta
in the recent years

presence | body-related | objects

ID39

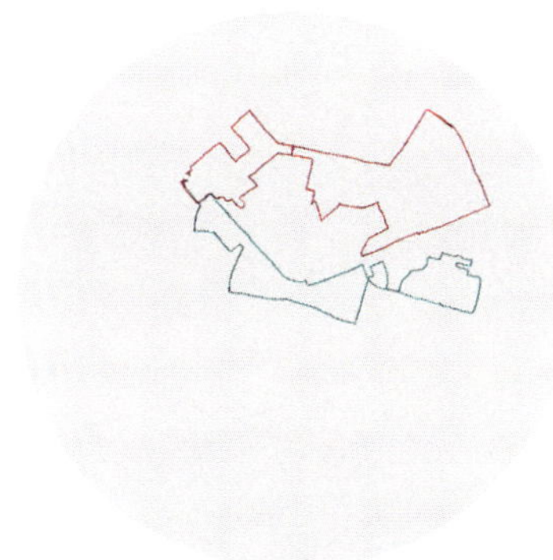

living in Nowa Huta from 1989 - 1993, and
since 1998

presence | body-related | NH layout

Figure 9 *Spread Routes (Map Type Atlas, Nowa Huta)*, digitalised maps, 100 × 100 cm each, marker on transparent paper (Simone Rueß, 2025, icons developed by Simone Rueß, digitally edited by Hanna Grodner, atlas layout by Hanna Grodner)

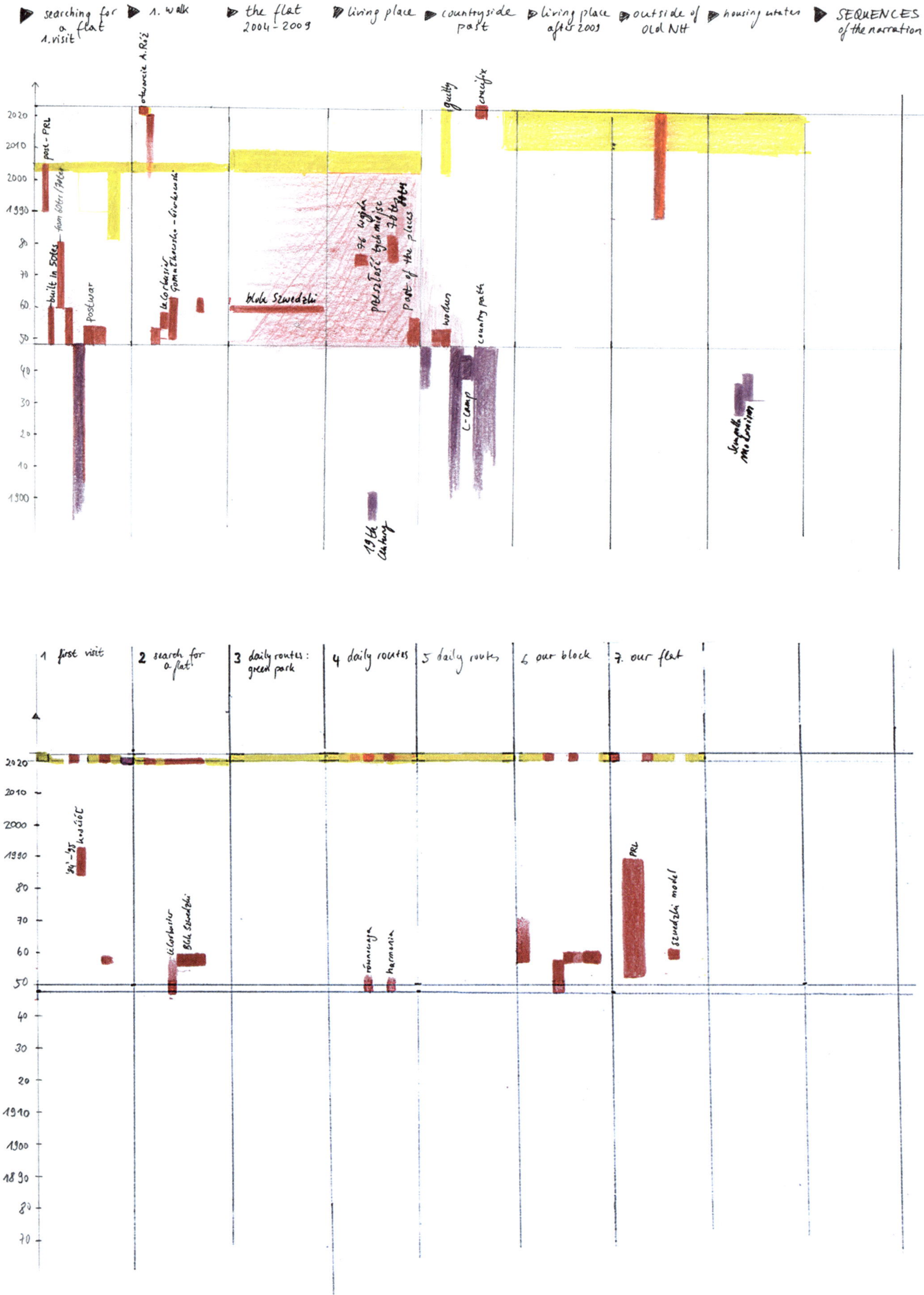

Figure 10 *Double Time Graphs (Nowa Huta)*. Left: ID04. Right: ID17, pencil and coloured pencil on paper, 21 × 29 cm each (Simone Rueß, 2024)

ID17	Sequence 1 **1st Visit**	Sequence 2 **Search for a flat**	Sequence 3 **Park Szwedzki, Vistula Meadows**	Sequence 4 **Daily Routes**	Sequence 5 **Other Daily Routes**	Sequence 6 **Residential Block Swedzki**	Sequence 7 **Individual Flat**
Temporality	Biographical time Historical time	Biographical time Historical time	Presence Biographical time Historical time	Presence Biographical time Historical time	Presence Biographical time	Presence Biographical time Historical time	Presence, Biographical time, Historical time
Route type	First visit to NH together with partner in order to visit a new antiquarian bookshop. Eldery local resident didn't know it yet (subject of gentrification)	Visiting NH for searching a flat with partner, explaining the environmental/architetural impression of Plac Centralny	Daily walk with dog, having ice cream	Routes along residential estates, green parks, pointing out places and objects of daily habits (pizzeria, theatre Łaznia Nowa/ Dom Utopii)	Intersection close to the own residential estate, Litfass column, Cafe, dogs' hairdresser, parks, other rare walks	Daily habits, experience of the home block	Daily habit, experience of the home flat
Content of the Sequence, Nowa Huta explanations	Park Szwedzki serves as a place to orientate on the map. Exit from tramway, striking church architecture, historically striking residential block and its first „slum" impression	Plac Centralny- a place where all streets concentrate around. Property/selling situation, atmosphere impression. Block Szwedzki: architectural-historical importance	Green park and Łaki Nowohuckie, place of best ice creams (subject of gentrification)	Characteristics in the different residential estates: order, harmony and chaos, meaning of green areas as an open space	Crossing as a bundling of information and movements, importance of the two parks close by for daily life	Architectural-historical characteristics of the layout and design of Block Szwedzki and its design in relation to the residential estate	Architectural design of the flats in PRL and the architectural-historical characteristics of Block Szwedzki
Drawn route character	Target	Target	Circle	Circle, target	Circle, target	Additive	Additive

LEGEND

biographical time
historical reference, time after 1948
historical reference, time before 1948, when Nowa Huta didn't yet exist
architectural reference to the Ideal City layout
nature reference
recent transformations

presence
biographical time
historical time

body-related
home-related
environmental-related
object-related
nature-related

residental estate
Nowa Huta layout
buildings
Łąki Nowohuckie
Zalew Nowohucki

circle
target
addative

Figure 11 *Table ID17 (Nowa Huta)*, table (Simone Rueß, 2024, icons developed by Simone Rueß, digitally edited by Hanna Grodner, table layout by Hanna Grodner)

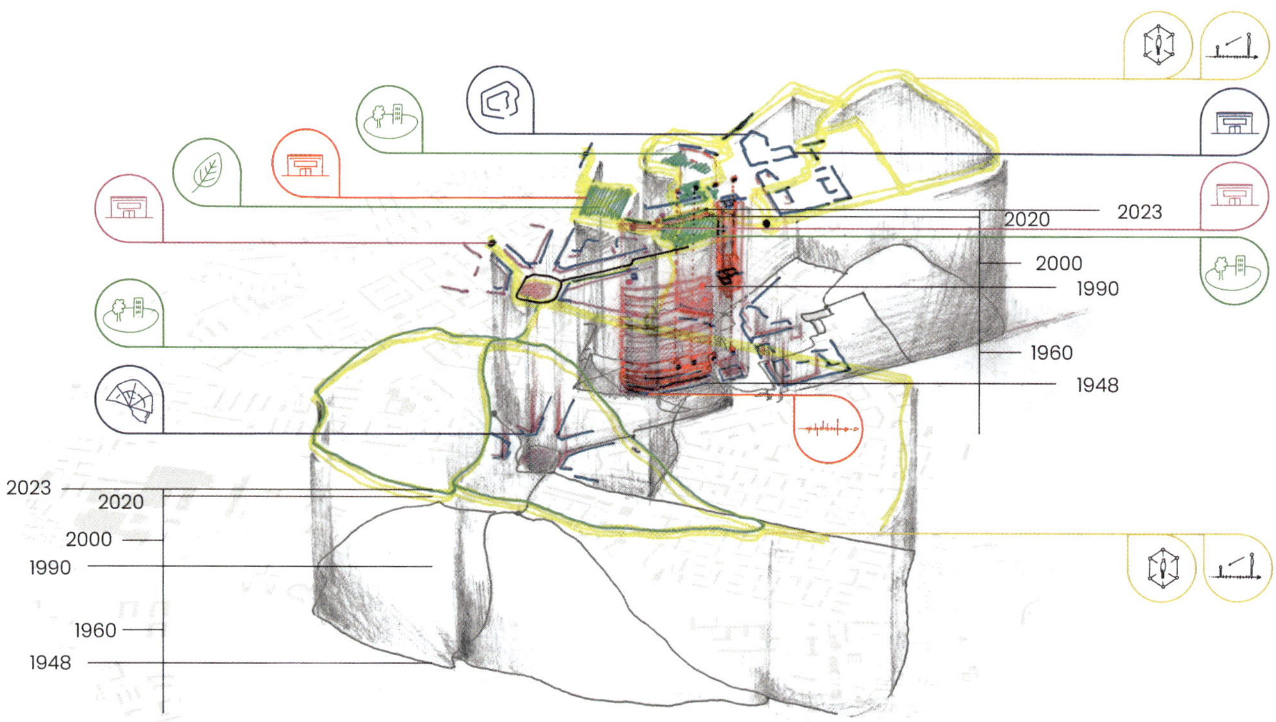

Figure 12 *Mental Timescapes (Nowa Huta)*. Above: ID04, below: ID17, pencil and coloured pencil on transparent papers, 21 × 29.7 cm each (Simone Rueß, 2024/25, digital illustration by Hanna Grodner and Simone Rueß, icons developed by Simone Rueß, digitally edited by Hanna Grodner)

environment but also to the actors' biographies and memories. Everyday routines can hardly be separated from references to their emotions and biographies. The city, too, cannot be read separately from its history—its 'biography'. Nowa Huta is 'inhabited by a multitude of different entities and personal and historical readings' (Foá et al. 2020, 138). With reference to Michel de Certeau, Phil Smith calls the processes 'by which the individual brings their story to a place' a 'Mythogeography'(Foá et al. 2020, 138). It is the intersection of both temporalities that emphasises the specificity of the city. But we have to ask how the present cultural writing of history impacts the residents' reflections. Memory changes over time, as does the cultural and collective narrative of history in societies.

Conclusion

The sensual tracing of the ground's texture, the tactile and embodied surveying of nonplaces, the physical connections with everyday life, the mapping of actors' movements in relation to their surroundings, and memory mapping in the city—all part of the multilayered project in Nowa Huta—relates to various types of mappings by artists such as Halcrow, Fassler, Alÿs, Emanuele, and Nold.

The flow of movement superimposed into simultaneity gives an idea of the creation of space over time—the so-called *movement spaces*. In the repetition of daily activities, I try to capture the continuity of forms that invisibly shape the spaces we inhabit. Movements change with time, but memories, perceptions, and imaginations also change over time. To get a glimpse of the relational structure of place, movement, and memories, I use a combination of artistic imaginary drawing and diagramming approaches. Visualised together, individuals' daily activities and personal histories—and the relationship of these to the urban landscape and its social and built histories—create an image of the complex interconnections of multiple temporalities and physical spaces.

The audiovisual installation *city/sound/scape,* in which Mazur and I accumulate our multiple mappings, reveals a sensual experience of interconnected temporalities and various spaces in Nowa Huta (Figure 1). Film images move across the walls, break off, and reappear elsewhere as the hand draws textures of movements projected onto the wall. Headphones transport sounds, voices, and noises, which blend into superimposed sound textures of the city. Video projections on the floor show drawing and gesticulating hands explaining their experiences in the city, while on the wall, the same lines—only without the city map—appear as an abstract animation in resonance with the sound textures. However, as (hi)stories and soundscapes change over time, we can only look at fragments of this complex temporal-spatial texture, giving us an idea—an imagination—of an ever-changing spatial multiplicity.

Acknowledgements

I would like to thank all of the interviewees in Nowa Huta, the team at Dom Utopii, and Teatr Łaźnia Nowa for all their support. I send heartfelt gratitude to Rafał Mazur for the wonderful interdisciplinary exchange and collaboration in *city/sound/scape.* Special thanks to the project curator Małgorzata Szydłowska and content coordinator Łukasz Trzciński, as well as to Dominik Stanisławski, Małgorzata Szymczyk-Karnasiewicz, Hanna Grodner, Aleksander Trafas and Tomasz Talerzek, for the great collaboration. Many thanks to the Foundation for Polish-German Cooperation and the City of Kraków for funding the project. Finally, a big thank you to Jamie-Scott Baxter, Séverine Marguin, and Vivien Sommer of the Hybrid Mapping Group, Eva Korte and deepani seth for all the theoretical exchange and reflections along the project.

Image Credits

Figure 1, 2, 4, 5, 8, 9, 10 © 2024 Simone Rueß and VG-Bild-Kunst
Figure 3: Lydia Harlcrow, 2019
Figure 4, 6: Images based on the performative and participatory research project *Daily Routes (Nowa Huta),* 2023, collaboration with fourty-seven participants during six days, part of *Archaeology of Everyday Life-New Topographies*, Teatr Łaźnia Nowa, Nowa Huta
Figure 7: Ella Emanuele, 2020/2022

References

Bergson, Henri. 1993. *Denken und Schöpferisches Werden: Aufsätze und Vorträge.* Hamburg: EVA Europäische Verlagsanstalt.

Emanuele, Ella, David Hunter, Zach Duer, and Simon Birch. 2024. 'dAnCing LiNes'. In *C&C 24: Proceedings of the 16th Conference on Creativity & Cognition*, 23–26 June 2024, pp. 614–618. https://doi.org/10.1145/3635636.3660502.

Foá, Maryclare, Jane Grisewood, Birgitta Hosea, and Carali McCall. 2020. *Performance Drawing: New Practices Since 1945.* New York and London: Bloomsbury.

Furján, Helene. 2009. 'Autonomous Worlds: The Works of Larissa Fassler'. In *Walking in Place.* Berlin. http://www.larissafassler.com/texts.html.

Gablowski, Birgit. 2013. 'Bergsons Appartment oder Räume als Chiffren zum Aspekt des Metaphysischen in Simone Rueß' Arbeit'. In *Appartement, Simone Rueß.* Karlsruhe: urbik.org.

Grünewald, Ursula. 2017. *Francis Alÿs. Die Stadt als Handlungsraum.* München: edition metzel.

Halcrow, Lydia. 2022. *A Thousand Intertwinings: An Exploration of Embodied Artistic Processes Made in Collaboration with an Estuarine Landscape and Its Vibrant Matter.* PhD thesis, Bath Spa University.

Kępiński, Kacper. n.d. 'Funktions on Nowa Huta after 89'. *Ideal City.* Accessed 2 September 2024: https://idealcity.pl/en/paths/functions-on-nowa-huta-after-89/.

Kern, Leslie. 2022. *Feminist City: Wie Frauen die Stadt erleben.* Translated by Emilia Gagalski. Münster: Unrast Verlag.

Kusiak, Joanna. 2012. 'Warsaw from Patelnia: Lessons in City Logic and Geometry'. In *Movement Space, Simone Rueß.* Warsaw: Fundacja 93 and Galeria le Guern.

Löw, Martina. 2016. *The Sociology of Space: Materiality, Social Structures and Action.* New York: Palgrave.

Lynch, Kevin. 1960. *The Image of the City.* Cambridge: M.I.T. Press.

Massey, Doreen B. 1994. *Space, Place and Gender.* Oxford: Blackwell and Polity Press.

Massey, Doreen B. 2005. *For Space.* London: Sage.

Rueß, Simone. 2023a. 'Drawing and (Re)Acting: The Creation of Movement Spaces'. *Drawing: Research, Theory, Practice* 8 (1): pp. 81–94. https://doi.org/10.1386/drtp_00107_1.

Rueß, Simone. 2023b. 'Nowa Huta (Hi)stories in Multiple Mapping Processes'. In *New Topographies of Nowa Huta*, edited by Trzciński, Łukasz. Kraków: Dom Utopia -Teatr Łaźnia Nowa.

Trzeciak, Katarzyna. n.d. 'Body'. *Ideal City.* https://idealcity.pl/en/paths/body/.

Weidenhaus, Gunter. 2015. *Soziale Raumzeit.* Berlin: Suhrkamp.

Dagmar Pelger

Time-Deep Mapping

Dis/continuities
in Another—
Commons-Based—
History of Urbanisation

The city as common space is not only contested and threatened, but it is also constantly being restored, preserved and renewed—through emancipatory movements and civil society action on the one hand, and through various modes of cooperative planning and institutionalised models of self-governance on the other. A history of the city from the perspective of the commons provides numerous examples and experiments in which urban spaces are kept open by means of communitisation as spatial commons. In order to support the integration of the self-empowered appropriation practices of common spaces on the one hand and their design principles on the other, this other story must be constantly retold in order to continue writing and producing it.

This research into another—commons-based—history of the city can be best conducted by mapping out timelines throughout the various epoques of urbanisation. With reference to medieval types of commons, modes of communal space production are traced in their historical continuity. From fifteenth-century beguinages to nineteenth-century one-kitchen houses, from nineteen-seventies squatting to contemporary housing projects in the syndicate model or model projects in self-managed communal cooperation, examples of urban renewal are examined as reproductive, use-oriented, feminist, subsistence-economic and non-exhaustive ideas of space and property.

The timeline—as a concept for tracing social movements, urban production of space as well as canonic historic events from the perspective of commoning—is used for visualising urbanisation as a (dis)continuation of ownership models from public to private, from common to club. This text will explore this concept.

For a Recollection of Examples in Urban Space Production as Commoning

In recent decades, the concept of the commons has received increasing attention both in theoretical discourse (Hardt and Negri 2009) and in the practice of urban social movements (Federici 2004; Harvey 2012) due to the growing interest of the financial sector in real estate and land. Questions about the preservation and production of commons as a countermodel to market orientation in urban development have become a central challenge for critical theory and the practice of urbanisation. However, the historical continuities and exemplary predecessors are not part of the dominant understanding of urban history, and only a few scholars refer to commons research or concepts when it comes to planning theory, urban development models, or practices of cooperative space production. The numerous forgotten, overwritten, or untold historical continuities and exemplary predecessors of the commons provide important references not only for the theoretical sharpening of the concept but also for its concrete implementation in planning and practice (De Angelis and Stavrides 2009).

To uncover the continuities and discontinuities, recalling urbanisation history from the perspective of the commons is urgent in order to contribute to the exchange and integration of collectivised spatial knowledge from the everyday practices of the commons, and professionalised spatial knowledge from the institutional commons. Recalling such an integrated commons-based perspective on history will help both theory and practice—theorists can weave the cases of commoning more systematically into the definitions of the commons and sharpen them on a spatial level, while practitioners can embed examples of commoning as precursors of cooperative planning for the appropriation of open space, securing living space, and self-organised working methods and operational models. The role of urban planning as a structuring mediator between the needs of those involved in the production of space requires both a theory to define common planning and a practice that makes the communalisation of spatial resources feasible.

However, to accomplish this, it is necessary to clarify not only how commons can be recognised and conceptualised in space but also how they differ from non-commons. Only then can they be traced as space-forming processes at different times and in different places. This iterative interplay

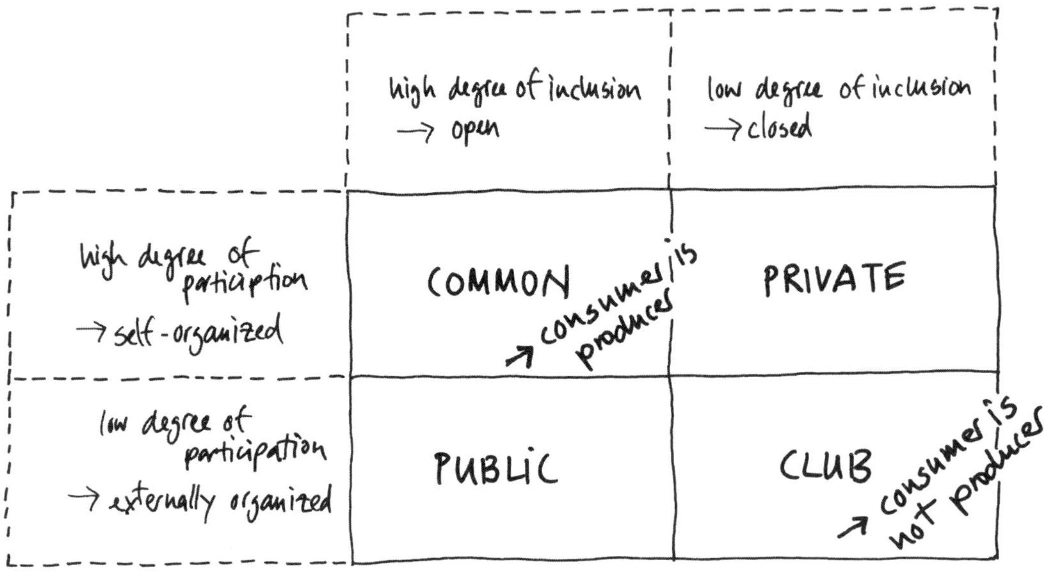

between a rereading of historical continuities and their retelling as the history, theory, and practice of commoning contributes to deepening and operationalising an emancipatory idea of the city as a commons itself (Lefebvre 1968; Stavrides 2016).

Research into these other commons-based—and thus different from the hegemonic—histories of the city calls for analytical and observation tools that enable the integration of time and processes into the spatial descriptions of practices of commoning. Mapping out timelines is an established technique in historical research that makes the process visually readable. However, it also reflects a linear understanding of progress that is deeply rooted in Western, patriarchal, and capitalist theory and thinking. Thus, how do we break with this linear logic without refusing the nature of temporality? How do we find a mode of timelining that helps shift the focus from the representation of the production of spaces within geographies and cartography to the representation of the production of spaces within timescapes and events as a processual reinterpretation of linearity? Timelining in various time-space trajectories throughout the waves of urbanisation is the experimental approach proposed as a tool here.

In this paper, only a selective contribution to this endeavour can be made, one that is formulated from a still very Western and therefore insufficiently inclusive point of view due to the places and events being examined. The aim of a continued retelling of urban history must be to broaden and, ideally, overcome this viewpoint by interrelating it with the many other histories of many other places. To shed additional light on the Eurocentric viewpoint will be the first step. To prepare for reflecting on the method of time-deep mapping, some general notes on spatial commons will help shift the perspective towards a commons-based history of urbanisation.

What Are Spatial Commons, and What Are They Not?

Our theoretical knowledge of the commons draws on two different fields of research: political and economic research into commons institutions (Ostrom and Ostrom 1977) and humanities and social science research into emancipatory commons practices and movements (De Cauter 2013; Federici 2004; Harvey 2012). To relate the characteristics by which commons can be recognised and distinguished from non-commons, sometimes contradictory

Figure 1 Four types of space diagram (Dagmar Pelger, 2022)

findings from both research fields must be linked together. From this, a series of definitions can be brought together on a spatial level and used as the basis for a historical consideration of spatial commons in urban and architectural history. The starting point of the commons is their joint production. Any process of commoning space can only take place if those who use and utilise spatial goods or resources are also those who determine the rules for this use through negotiation (Ostrom 1990). However, only when spatial resources and the negotiation of rules are potentially accessible to all can the commons open up as a shared space in the sense of emancipatory appropriation and thus fulfil the description of the commons as a place of solidary relationships (Federici 2004).

Spatial commons, as a sphere beyond the public and the private, are determined, in particular, through sharing the yields from the respective use of a spatial resource. As long as the material, immaterial, social, or cultural yields—for example, a particular spatial commons for assembly (De Cauter 2014)—are shared among all and are not extracted by a few, the resource space itself—for example, the city as a universal spatial commons in which the assembly takes place (De Cauter 2013)—is also preserved and maintained by returning the surplus for the use of the city as a whole. Commons thus become legible as reproductively organised spatial systems.

From the four types of goods diagram established in political science (Ostrom and Ostrom 1977), we can also derive four types of spaces that make up the city: common, public, private, and club spaces. When translating this diagram into a diagram of four spatial types, it can describe not only the different ways of dealing with spatial resources—the yields, and the different degrees of accessibility—but also the different concepts of ownership, which reflect that when a higher degree of inclusion and self-governance prevails or can be established in a place or a process, it leads to a more comprehensive common good (Figure 1). This categorisation makes it possible to consider spatial commons as distinct from non-commons, such as public, private, or club spaces.

Time-Deep Mapping as a Multilinear Matrix of Places and Events

Our imagination and institutionalisation of ownership are in constant flux. Repeated shifts and transformations cause continuously new understandings not only of the world and the ongoing processes of urbanisation but also of the concept of ownership (von Redecker 2016). Initiated by various events in various places by various people, these transformations with reference to the city are interpreted, described, and written down as urban history—with the latent notions of ownership being inscribed into them. All written histories, of which the Eurocentric is only one, cause conditioning and, in turn, condition the way in which space is described, conceived, produced, represented, and conceptualised. Each of these histories consists of countless trajectories that run parallel, intertwined, and simultaneously next to each other, as well as around the chronologically countable events through time. Massey (2005) framed the multiplicities of simultaneous trajectories as time-space.

In the case of the dominant hegemonic understanding of Northwestern Europe, a strict hierarchy of certain trajectories and a strict division in time frames between major events characterise a canonical linear story written down by a White and patriarchally conditioned elite, certainly within urban planning and design. Critical theory, feminist and queer philosophy, and postcolonial studies are the latest discursive forces challenging this hegemonic understanding of history by elaborating on and uncovering counter- and repressed trajectories. To integrate these perspectives and discover more about suppressed, hidden, and wiped-out histories, a closer study in the urban history literature must include the influences of commoning-based action on a seemingly linear path along progress and growth that dominates the view on urban development and renewal.

For this, we need to reread the very well-known and almost-forgotten examples and stories in urban history literature. The known examples are often framed in the dichotomy of public versus private modes of ownership or commission and use. They need to be reinterpreted with the theoretical knowledge about commoning as a collective mode of space production beyond public and private. In addition, the almost-forgotten examples in urban history literature need to be retold, reintegrated, and reinterpreted as a legitimate part of the history of urbanisation.

As we have learned from Massey, both perspectives—the temporal and the spatial—are inseparable layers within a time-space relation that constitutes our understanding of the world—and thus the urban. An intermediate step in distinguishing the spatial *simultaneities* within processes of urban space production from temporal *trajectories* that create continuities and discontinuities between events within these processes might help contextualise and argue for the choices being made while elaborating another understanding of histories as time-space structures.

The methodological proposition here is to reread the various simultaneous trajectories as a multiplicity of conflicting movements in which the model of two use-based spatial types (common and club) and two territory-based spatial types (public and private) is repeatedly confronted and in which a commons-based notion of ownership and space production is displaced or opposed and, at certain times, strengthened and renewed (Figure 6). A certain colour code is applied for first interpreting the types as modes of ownership within which a certain project, example, phenomenon, plan, or spatial production has taken place. In the case of the presented experiment, red is used for common, blue for public, purple for club, and grey for private ownership. The rereading of places and events in history through this colour code leads to a cartography of changing institutionalisations and imaginaries of ownership through time. Along the conception of the temporal-spatial, wave-like movements of these changes in ownership relations and space production, the practice and concept of commons coloured in red becomes readable as a phenomenon—in struggle, conflict, and negotiation with the purple club and the binary concept of the blue public and the grey private conditions—of an inclusive, emancipatory, and collective mode of ownership that continuously recurs in various times and spaces. This is seen as a red thread that is only slowly revealed in current urban commons discourses.

The colour code, and thus the notion of ownership, can appear on four levels in the time-deep mapping of urbanisation. The chart of gradients of changing ownership as sociopolitical movements that are contradictory and simultaneous with the events associated with them is the underlying structure of the mapping. On the second layer, a selection of trajectories stretches out with the same colour code, but one based on continuities within the movements. They can refer to architectural projects as well as urban political interventions, urbanistic paradigms, or spatial phenomena related to social struggles or planning laws. These trajectories can be an argument for a certain dominant concept of ownership in a certain period of time or place, but they can also introduce a counternarrative to the dominant movement. A third layer in which the colour code is applied represents the categorisation of single projects, plans, buildings, urban areas, spaces, places, concepts, or utopias. They may appear as embedded in trajectories or movements, as stabilising carriers, or as singular events.

Through a combination of sphere-like drawing codes, (poly)line-shaped elements, and punctual markers, a spatially structured narration along a linear performance of passing years opens up a field of events within a time-deep chart. In the following, we will zoom in on the time-deep mapping and focus on certain time frames of the transformations between what is defined as medieval times and Early Modern times. A much broader view would be necessary but can also be interpreted by the reader when wandering through the chart.

Figure 2 Höttinger Alm (near Innsbruck), as an example of cooperative food production (Innsbrucker Stadtarchiv, n.d.)

Contradictory Simultaneities in Steadily Changing Notions of Ownership

Not long after the first accounts of urbanisation some 10,000 years ago, the spatial order in Northern Europe had a re-emergence of spatial types of commonings as a social model tolerated and sometimes even institutionalised by the dominant political elite. After the collapse of the power structures of the Roman Empire, the dichotomy of public versus private vanished around the Mediterranean Sea and was successively replaced in large parts of Europe by a pyramidal order of social and economic dependencies dominated by religious and aristocratic elites and based on serfdom, in continuation of ancient slavery.

Until the beginning of colonialism, this period, which is referred to as the feudalism of the Middle Ages, can be regarded—with respect to the focus on four types of space production and ownership—as the spatialisation of clerical and noble institutions in the shape of episcopal cities or princely states, along with the associated infrastructures of serfdom. Embedded in this nested system of club-like spatial entities of fortified towns around cathedrals or palaces, the commons re-emerged as a partly institutionalised subsistence economy. Through the appropriation of woods, meadows, or pastures by the peasantry or serfs, we can read the commons as a self-help system or appeasement measure by the clerical or noble landlords who owned the land, the resources, and the serfs to cultivate a slave-like dependency. However, in continuation of older traditions, the social practices established in the commons, the common land, woods, pastures, alps, and meadows were not only practised, given further, and elaborated (e.g., in the cooperative systems to organise the alp high up in the mountains) but were also implemented in urban contexts in the form of cooperative institutions, such as guilds and cooperatives, for the self-organisation of craftsmen and merchants.

A similarly emancipatory approach of independence from clerical and noble power structures can be read in the initiation of independent or free towns in Northern Europe and Northern Italy. Although access to these early examples of civic self-governance and a capitalist profit economy was also limited, they nonetheless reflect some features of the commons in that they removed religious or aristocratic rule and replaced it with a mercantile civic power structure that enabled social advancement through economic success.

The beginning of what is conceptualised as modernity—describing the rise of capitalist economics and politics—was marked by the start of colonisation by European empires and the subsequent successive disempowerment of the clerical and aristocratic estate-based society. This transition was accompanied by a shift in power in favour of the urban merchant class, which asserted its claim to exclusive access to spatial resources through the enclosure of common land, common practices, and commoners' self-regulation. This went along with the transformations of the institutions of ownership. The prospering civic upper class or urban bourgeoisie successively asserted private—and sometimes public—property titles to formerly unparcelled common land, with recourse to Roman law, which undermined the inclusionary and exclusionary modes of use in the common- and club-like spatial condition based on negotiation and domination. This reinstatement of the antique dichotomy of public and private accompanied implementations of municipal, state, and nation-state institutions, resembling the governance models of the patrician houses of the late medieval independent free towns. This formed the basis for enforcing an accumulation and capital-based extractivist urbanisation not only in Europe but also in the dominated territories on other continents.

Still, some commons institutions sustained the transformation towards the modern capitalist urbanisation processes. We find cooperative structures in urban developments, practices of communing, and subsistence economy in rural economies, as well as shared labour in the house, yard, and family as conventionalised formed of self-organisation, care, self-help, and even juridical systems of common ownership, and so on. Many of them last until today.

Figure 3 Freistaat Barackia (Berlin), as an example of self-organised housing provision (L. Loeffler, 1872)

In the context of the expanding extraction of resources, colonialism, industrialisation, and capital accumulation in the aftermath of peasants' revolts, civil revolutions, and monarchist counterrevolutions, the binary of public (state) and private (economic) ownership models were challenged during the economic boost of state-supported liberalism (and deregulation) in the nineteenth century. New institutions of ownership, management, governance, and operational and spatial systems seemed to deliver updates to the nested systems developed in medieval times. The commons and clubs experienced a kind of revival during liberalism and the new urban expansions that came along with it. While banking systems, stock corporations, and limited liability companies were developed and expanded with state support to manage and exploit natural, cultural, social, and urban resources under the legal protection of private (commercial) law, the accumulation of profits, and the speculation that came with it, club institutions and spatial systems were established and could only be maintained through shortened accessibility to resources—including housing within rapidly urbanised areas around the European major cities with expanding industrialisation.

In counterreaction and as a parallel trajectory, we can read the rising engagement of early socialist theorists and philanthropists with cooperative organisation. Some of the more known examples on which various urban developments are based include Charles Fourier's utopian concept of the *phalanstère*, which was based on a building type similar to the Palace of Versailles that was organised as a social palace for housing in communes instead of families (around 1830) (Bebel 1890), or later, Ebenezer Howard's 1889 vision of the garden city as a radial urban layout of a neighbourhood in cooperative ownership (Howard 1902). They both implicitly recuperate commoners' traditions that never stopped being practised, despite all the violent waves of enclosures. Little has been reported or documented on the everyday actions, oral histories, traditions, and conventions of solidary self-organisation on which theorists based their work—in comparison to the legends of Fourier and Howard, who had the resources to publish and write their own histories. However, we can read the struggles from below in between the lines of the theories, design proposals, and development projects, and understand them as recuperation of the traditional practices in the commons (Federici 2004; Linebaugh 2008).

Trajectories of Spatial Appropriation as Practices of Commoning

The above-described excerpts—from a matrix that started much earlier and ended much later with an open future—lay out only a small part of the experiment for countering the hegemonic narrative of a linearity of events that frame urban history. The matrix, as an excerpt in itself, is meant to serve as a testbed for a commons-based view of the dynamic simultaneous multiplicities of urbanisation. The following explores this testbed as a time-deep context for uncovering the various spatial structures and processes of urbanisation as multiple trajectories of a commons-based production of the urban. To do so, a selection of trajectories is formulated to uncover the dis/continuities of these events and places through time to help us read urbanisation from a commons perspective.

Starting from the traditional commons of feudalism, we can best follow the traces of commoning in an open-space condition. There, commoning practices appear in various forms of the rural subsistence economy and thereby form reproductive spatial systems. As such, the alpine pasture, as a communalised seasonal farm high up in the mountains, was one of the earliest types of commons described (Figure 2). The extreme location high up in the mountains required a high degree of self-organisation that developed into cooperative models, some of which still exist today. However, the archetype of the feudal commons is the forest clearing, in which the peasantry collected firewood and, in certain contexts, gained the official right to do so. Other traditional types of commons, which had a central social function in the village community as pastures, fish ponds, vegetable gardens, or fairgrounds, could already be interpreted in preantique urban forms as unbuilt circular open spaces in the ring-shaped settlement of Nebelivka or the nested courtyard system in the clustered city of Çatalhöyük (Graeber and Wengrow 2021). In the shape of occupied squares, village greens, parks, and urban voids, common space as an accessible resource is still present today.

The rural subsistence institutions of commoning were enclosed and privatised from the beginning of colonialism on and mostly vanished as institutions throughout industrialisation. Some local types remained in collectivised hands as hut pastures, wandering pastures, or meadows, while some were gradually transformed into public parks or village squares. In today's Europe, some unenclosed commons can still be found in the Northern Italian foothills of the Alps, as forest areas in common ownership of the Usi Civici (Elzenbaumer 2020), or in the middle of London's urban fabric as publicly accessible common lands. However, even if commons are preserved as spatial figures, they could only be distinguished as places of commoning on the basis of inclusive use. The northern areas of London, which still consist of large estates on private property, are characterised by numerous landscaped and fenced little squares. They may appear as commons embedded in the street space, along which townhouses line up around them. However, they differ greatly in terms of use and ownership from actual common land, as they are exclusively accessible and exploitable by only the neighbouring property owners in terms of both open-space recreational yield and contribution to property price increases they thus form club goods with high revenue growth.

In rapidly growing industrial cities, informal settlements in the peripheries were installed due to the extreme housing shortage caused by rent price increases. These self-built and occupation practices can be read as an opposition to the enclosure of spatial resource housing, which became a major asset of speculative urban development in the liberalised speculation market. During the large-scale demolition measures by Napoleon III and Haussmann in Paris, which at the time still had a medieval structure, a self-built settlement with up to 30,000 residents was created in La Zone (nonconstructible) around Paris. The violent transformation of the city centre took place over many decades and dissolved the street space from the small-angled common condition of alleys and market stalls into the straight-lined club condition of boulevards and salons. This urban renewal policy should be considered another enclosure of common space. At the same time, in Berlin, the self-proclaimed Free State of Barackia (Figure 3) was installed outside the city gates, starting from Landsberger Tor and Kottbusser Tor, as a proletarian self-built housing estate with self-provision gardens on the plots for investor-developed tenements intended for Hobrecht's urban expansion (Krasny 2012). In their quality of self-organisation, empowerment, and utopian outreach, both examples—despite being embedded in politically different situations—show references to a still ongoing tradition of cooperative space production that merges the spatial practice of the traditional commons with the modern condition of the industrial metropolis, leading to the coexistence of urban types that the hegemonic narrative of the city separates in time and space from each other (Massey 2005).

Likewise, principles of occupation and self-construction were reflected in the practice of the Open Space Committee, a civil society mapping project formed around social ecologist Patrick Geddes in Edinburgh. Derelict land was mapped to prevent it from being sold and then appropriated by the neighbours for gardening, play, and recreational activities. Stairs and fences were constructed to make the plots accessible and to manifest their claims of common use. In a similar way, the allotment garden movement was based on a self-help initiative for horticultural self-sufficiency with self-help components for cultivating open spaces. Meanwhile, the principle of allotment gardens has diversified to include various mixtures of common- and club-like spaces,

Figure 4 Gecekondu (Berlin), as an example of a self-built protest camp and community centre (Beke Bücking, 2022)

whereby the fundamental nonprofit organisation clearly assigns them closer to the concept of commons.

The resumption of self-sufficient and communal gardening as urban gardening movements from Detroit to Saõ Paulo and Berlin, as well as the occupations of squares in Athens, Istanbul, Cairo, New York, and Madrid following the financial crises of 2008, bears witness to a wave of scarcity of spatial resources through their resistance to enclosure. In addition, they continue the narrative of the ongoing commoning described above. The occupations have given rise to urban open spaces in communalised ownership, such as Navarinou Park in Exarchia or Gecekondu in Kreuzberg (Figure 4), in which not only the spatial yields and benefits but also the burdens of the scarce availability of resources are communalised as negative commons (Harvey 2012). They show that the changing conditions for the reproduction of common space are constantly giving rise to new variants of inclusive use of resources and the self-managed invention of codes and conventions of commoning.

These few selected deconstructive intermediate steps help distinguish the spatial *simultaneities* within processes of urban space production from the temporal *trajectories* that create continuities and discontinuities between events within those processes. They make legible the choices being made in elaborating a certain understanding about what has happened so far (Figure 5). A perspective adopting four types of space and ownership was used to contextualise the view of spatial simultaneities, and the focus on common ownership in open space appropriation was chosen to argue for the selection of the temporal trajectories presented.

Conceptualising Potential Futures as Imaginaries of Common Ownership

One of the difficulties of reinterpreting and reimagining urban history and urban future as a dynamic and non-linear structure is the lack of documentation and representation of the modes of urban spatial production through commoning. They have been largely erased from the hegemonic narrative or overwritten by the dominant narration systems—of the state and market, in the case of here and now. The European–Western linear urban history concept is based on a patriarchal, colonialist, and capitalist narrative that insists on a static relationship between the public and private. It proclaims that any solidary common-like and extractivist club-like use (whether ownership, or production of space) is hidden in the dark of the private sphere. This construction continues to shape our notions of society, space, and property and, with few exceptions, the educational, practice, and research institutions of architects and planners.

For a reproductive, solidarity-based, and resource-sensitive renewal of the city that is oriented towards the common good, the largely hidden and forgotten trajectories of the commons as practices of self-help, subsistence, and emancipatory solidarity must be exposed again and again. Writing history in a different way is all the more urgent in the context of current ecological, social, and political crises and represents a central task of urban research. With this comes the claim and search for different methods and techniques.

Time-deep mapping, as interpreted and described above, is only one possible narrative among many other dynamic simultaneous multiplicities overlaid with a selection of multiple trajectories. The empowering potential in both explorative layers—the matrix based on four types of ownership and the trajectories of commoning within—lies in the claim for the active involvement of many authors and participants in the constitution of remembered histories and open futures. By mirroring those events and places that we find worth remembering today into an open-future tomorrow, we decide its constitution every day.

The powerful tools of cartography and histography must be reappropriated, shared, and opened up to a wider and more diverse authorship. With a focus on the definition of commons as an inclusive mode of spatial production, the integration of new tools and other dimensions of time-space offers a perspective on urbanisation that challenges our perception of ownership, agency, and situatedness in space and time. Other ways of political thinking and acting beyond the extractive market logic will occur out of the growing pressure on resources. Planning and design as projective disciplines that are conditioned through political decision-making for the common good will face the need for and emergence of tools, instruments, and methods that are capable of reproductive and nonextractive thinking and functioning.

The potential in continuing to reference time-deep mapping with factual events that also provide orientation in the dominant history writing is the reinterpretation of that reference frame within the adapted setting of the matrix. A field with four categories of spatial types and ownership is stretched out in the y direction. In this field, four, eight, sixteen, thirty-two, sixty-four ... countless and multiple dynamic simultaneities of time-space trajectories can be relocated, and they intertwine, interrupt, emerge, dissolve, expand, or start looping with and around the official events of dominant history writing.

While highlighting the perspective of commoning in such an approach, the selection of those elements of collective memories—similar to the techniques of mapping—comes to the foreground and refers to a solidary, reproductive, radical democratic, sustainable, inclusive, and self-determined way of urban space constitution. By rewriting those recollected trajectories and embedding them within an ownership-based matrix of simultaneities, we not only re-collect what has happened so far but also conceptualise, imagine, and argue what will happen, with alternative futures of spatial commons as a basis for decision-making. This might be a fruitful ground for continuing with commons planning from below, not only in politics and science but also in urban and everyday planning and design.

Image Credits

Figure 1: Dagmar Pelger, 2022
Figure 2: Much Heiss (from the collection of the Stadtarchiv Innsbruck)
Figure 3: 'Die Barackenstadt von Berlin. Nach der Natur aufgenommen von L. Loeffler'; *Die Gartenlaube*, 1872
Figure 4: Beke Bücking, 2022
Figure 5: Dagmar Pelger, 2024
Figure 6: Dagmar Pelger with Beke Bücking and Polly Bruchlos, 2024

References

Bebel, August. 1890. *Charles Fourier. Sein Leben und seine Theorien*. Stuttgart: Dietz.

De Angelis, Massimo, and Stavros Stravrides. 2010. 'Beyond Markets or States: Commoning as Collective Practice (a Public Interview)'. *An Architektur* (23): pp. 3–27.

De Cauter, Lieven. 2013. 'Common Places: Preliminary Notes on the (Spatial) Commons'. *De Wereld Morgen community blog*. Available at: https://www.dewereldmorgen.be/community/common-places-preliminary-notes-on-the-spatial-commons/.

De Cauter, Lieven. 2014. 'Common Places: Theses on the Commons', *Depression Era texts*. Available at: https://depressionera.gr/lieven-de-cauter-i.

Elzenbaumer, Bianca. 2020. 'Community Economies, A Practice Exchange: 7-9 June 2019, Vallagarina, Italy'. *Alpine Community Economics Laboratory Snapshot Journal* (1). Available at: https://hdl.handle.net/10863/17124.

Federici, Silvia. 2004. *Caliban and the Witch: Women, The Body and Primitive Accumulation.* New York: Autonomedia.

Graeber, David, and David Wengrow. 2021. *The Dawn of Everything: A New History of Humanity*. London: Penguin.

Hardt, Michael and Antonio Negri. 2009. *Commonwealth*. Cambridge: Harvard University Press.

Harvey, David. 2012. *Rebel Cities: From the Right to the City to the Urban Revolution*. London: Verso.

Howard, Ebenezer. 1902. *Garden Cities of Tomorrow*. London: S. Sonnenschein & Co.

Krasny, Elke. 2012. *Hands-On Urbanism 1850–2012: vom Recht auf Grün*. Wien: Turia+ Kant.

Lefebvre, Henri. 2016. *Das Recht Auf Stadt*. Hamburg: Nautilus Flugschrift. First published 1968 *Le Droit à la Ville* by Anthropos, Paris.

Linebaugh, Peter. 2008. *The Magna Carta Manifesto: Liberties and Commons for All*. Berkely: University of California.

Massey, Doreen. 2005. *For Space*. Thousand Oaks: SAGE Publications.

Ostrom, Elinor. 1990. *Governing the Commons. The Evolution of Institutions for Collective Actions*. Cambridge: Cambridge University Press

Ostrom, Elinor, and Vincent Ostrom. 1977. *Public Goods and Public Choices*. Workshop in political theory and analysis. Indiana University.

Pelger, Dagmar. 2022. *Spatial Commons. Zur Vergemeinschaftung urbaner Räume*. Hamburg: Adocs Verlag und Produktion.

Stavrides, Stavros. 2016. *Common Space: The City as Commons*. London: Zed Books.

von Redecker, Eva. 2016. *Interview in 'Formen des Eigentums'*. Humbolt-Universität. YouTube. August 3, 2018. Video, 3:59. Available at: https://www.youtube.com/results?search_query=eva+von+redecker+teil+1.

Biographies

Amer, Qusay is an architect, researcher and doctoral candidate at the Chair of International Urbanism and Design (Habitat Unit) at the Institute for Architecture of Technische Universität Berlin, Germany. He is a research assistant at the subproject C08 'Architectures of Asylum: Circulation of Governance Approaches, Planning Knowledge, Design Practices, and Materialities' at the Collaborative Research Centre 'Re-Figuration of Spaces' (CRC 1265) at Technische Universität Berlin. He studied architecture at the Brandenburg University of Technology. His research focuses on the production of physical spaces for and by migration, refugees' urban agency, and the agency of urban poor communities, as well as the translocal (urban) spatial governance.

Baxter, Jamie-Scott, Dr.-Ing., is a senior design researcher at the Chair for Urban Design and Urbanisation, Technische Universität Berlin. He is director of 'Planetary Tactics for Cohabitation' lab within the project 'Re-Scaling Global Health. Human Health and Multispecies Cohabitation on an Urban Planet'. In this constellation, he jointly coordinates the sub-project 'The Health Effects of Biophilic Urbanisms: Planning with Pathogens in the United Kingdom'. He also leads 'Planetability', a project that links planetary health, planteary thinking and multispecies cohabitation work between Brazil and Berlin. Jamie is an associted member of the CRC 1265 'Refigurations of Spaces' in Berlin and joint executive editor at the journal *Architecture and Culture*.

Carr, Lili is an architect and landscape designer with an interest in the soils, waters, and ecologies of urban environments. She received her BA from the University of Oxford, her diploma in architecture from the Architectural Association, and her masters in landscape architecture from the University of Edinburgh. She is currently a STARTS/European Commission supported artist-in-residence at the BOKU Institute for Hydraulic Engineering and River Research in Vienna, Austria, where she is exploring the material, social and ecological effects of the city's energy demand and attendant infrastructures on the floodplains and riparian zones of the Upper Danube.

Ceola, Francesca is a researcher and PhD candidate at the Habitat Unit, Institute of Architecture of the Technische Universität Berlin, Germany. She is researcher for the project 'Architectures of Asylum II: Circulation of Governance Approaches, Planning Knowledge, Design Practices, and Materialities' at the Collaborative Research Centre 'Re-Figuration of Spaces' (CRC 1265), also at the Technische Universität Berlin. She studied Sustainable Development at the University of Edinburgh (Scotland, UK), and Human Geography at the Universidade de Lisboa (Portugal).
Her research explores mapping the intersections between forced displacement, makeshift urbanisms, and socio-political ecologies in shifting landscapes in Lagos (Nigeria). She is involved in interdisciplinary creative collaborations' approaches to research—through film making, performance, counter mapping, and dance.

Coffey, Victoria Baskin is a visual anthropologist and media producer with an enduring interest in the ways images make the world. Their work spans diverse cultural contexts, fostering collaborations with communities, artists, and scholars to examine how co-creative research and expressions nurture resilience and renewal. Victoria is the Visual Editor of the digital project *Feral Atlas* and co-curated the project for exhibition around the world. They are currently a member of the intercultural collective Miyarrka Media while also pursuing a PhD project in collaboration with Shanthi Muniswamy. Supported by James Cook University, Aarhus University, and IIT Madras, Shanthi and Victoria explore the digital-visual practices among transgender and non-binary communities in Southern India.

Constantin, Lus is an independent multidisciplinary artist who combines his knowledge of architecture, art and digital archives to create immersive media experiences. Previously, he worked with architectural practices in London and Berlin where his collaborative work has been published in national media. Since 2010, he manages the reappraisal and digitisation of the artist estate of the German sculptor Dieter Finke. Currently, his interests include scripting architecture to transform the digital landscape, questioning how we interact with mixed reality environments. His artistic research investigates challenges of bringing large data sets on to the blockchain and making them accessible in Meterverse experiences. In March 2023 his virtual architecture work was showcased in Monaverse.

Deger, Jennifer is professor of Digital Humanities and Director of the Centre for Creative Futures at Charles Darwin University, Australia. She is chief investigator on the Yolŋu-led Australian Research Council projects 'Raŋipuy: a Yolngu Digital Arts of Renewal' and 'Caring for Cosmologies: Making Living Maps for West Miyarrka'. Trained in communications and anthropology, she is a founding member of the award-winning arts collective Miyarrka Media and co-curator of the digital project *Feral Atlas*. Moving between film, installation and intermedial writing, her work seeks to nurture new zones of more-than-human expression and co-creativity.

Fry, Tony, PhD., is an award-winning designer, cultural theorist, educator and author. He is adjunct professor, Architecture and Design, University of Tasmania, visiting professor, University of Ibaguè (Colombia), and the principal of The Studio at the Edge of the World, Tasmania. His research interests include climate futures, post-sustainability, ontological design, cosmotechnics, decoloniality, politics and conflict. Tony has published twenty books. He is currently working on two projects: the relocation of communities at risk from climate change impacts in Australia and the communication of the existential condition of 'living in tension' with Latin American collaborators. Tony has two books forthcoming. *Contrapractice* with Dulmini Perera (Bauhaus Universität Weimar), and *Political Breakou*t, his second book on a *New Political Imagination*. Both books are due to be published in late 2025, early 2026.

Heinrich, Anna Juliane, Dr.-Ing., is senior researcher and lecturer at the Chair of Urban Design and Urban Development at the Institute of Urban and Regional Planning of Technische Universität Berlin, Germany. She is principal investigator of the DFG-funded research project 'The Spatial Knowledge of Young Adults: The Constitution of Online, Offline and Hybrid Spaces' and Co-Head of the graduate school of the Collaborative Research Centre 'Re-Figuration of Spaces' (CRC 1265) at Technische Universität Berlin. She studied spatial planning at TU Dortmund and KTH Stockholm and did her doctorate at the Department of Urban and Regional Planning of TU Berlin. Her research covers spatial theory, planning processes and cultures, spatialities of childhoods and youths, participation and co-creation in urban development, as well as methodologies and methods of research in planning and design.

Lee, Jae-Young has been affiliated with the Leibniz Institute for Research on Society and Space and the Collaborative Research Centre 'Re-Figuration of Spaces' (CRC 1265) at Technische Universität Berlin since 2022 within the project 'Peripheralised Rural Areas: Digitalisation and Constructions of Space'. As an architect, she has been part of the practice Gyaw Gyaw at the Thai–Burma border since 2016, working on communitarian architectures. Her research interest is in territorial dynamics and grassroots practices that shape contemporary peripheries.

Marguin, Séverine, Dr., is a cultural sociologist and deputy professor for 'Spatial Sociology and Interdisciplinary Spatial Research' at HafenCity University Hamburg. In this role she is involved in the Grako 'Urban future-making: Professional agency across time and scale' at the HCU. Her interdisciplinary teaching and research focus on cultural fields, knowledge production, sociology of space, natureculture relationships, experimental and design-based methods. In her current DFG funded research project 'Afronovelas: Spatial Stories and Production Regime in West-African Soaps' within the Collaborative Research Centre 'Re-Figuration of Spaces' (CRC 1265) at the Technische Universität Berlin, she develops a sociological perspective on the fictionalisation of spaces. She is joint executive editor at the British Journal *Architecture and Culture*.

Million, Angela, Dr.-Ing., is professor and Head of the Chair of Urban Design and Urban Development at the Technische Universität of Berlin as well as director of the DAAD-exceed Global Center on Spatial Methods for Urban Sustainability. She holds an adjunct professorship with Michigan State University. She is a principal investigator of the research project 'The Spatial Knowledge of Young Adults: The Constitution of Online, Offline, and Hybrid Spaces' at the Collaborative Research Centre 'Re-Figuration of Spaces' (CRC 1265) at Technische Universität Berlin, as well as the principal investigator of the project 'Well-being in Socio-Spatial Contexts: Intersectional Perspectives on Children's Experiences at Non-School Learning Sites' (WIKKI). Her current research explores educational landscapes, geographies of children and youth, multifunctionality of urban infrastructure and neuro-urbanism and well-being.

Nobre, Ana Luiza, Ph.D. in History, is professor of theory and history of architecture at the Department of Architecture and Urbanism of the Pontifical Catholic University of Rio de Janeiro. She is researcher of the Brazilian National Council for Scientific and Technological Development, researcher at the Center for Studies on Contemporary Spatialities of the Institute of Architecture and Urbanism at the University of São Paulo, member of the International Committee of Architectural Critics, and Co-Coordinator of the *Ground Atlas* project (groundatlas.org). She studied Architecture at the Federal University of Rio de Janeiro and did her doctorate in history at the Pontifical Catholic University of Rio de Janeiro. Her research moves across architecture, urban studies, critical history, arts, and cartography and is inspired by feminist and decolonial approaches.

Önen, Simten acquired a bachelor's degree in architecture from the Middle East Technical University in Ankara in 2022. Her diploma studies focused on urban development projects on the coastline of Istanbul. Since 2022 she is a Master Urban Design student at the Technische Universität of Berlin and was research tutor within the project 'Well-being in Socio-Spatial Contexts: Intersectional Perspectives on Children's Experiences at Non-School Learning Sites (WIKK*I)'.

Pelger, Dagmar, Dr.-Ing., is an architect and currently visiting professor for Sustainable Cities and Communities at the ASL department, University of Kassel. She most recently taught as a guest lecturer for urban design at UdK Berlin and TU Berlin. Her research focuses on spatial commons as urban spatial systems of collectivisation, on models of cooperative planning between municipalities and civil society, and on mapping as interdisciplinary and participatory research and design method. In 2025 her dissertation *Spatial Commons. On the Communalisation of Urban Space* will be published in English (adocs Hamburg). Since 2017 she has been part of coopdisco in Berlin, an architecture and planning cooperative committed to urban development for the common good.

Rekacewicz, Philippe is a geographer, cartographer, and information designer. He was an editor at *Le Monde diplomatique* (1989–2012), co-leading the production of geopolitical atlases (2002–2012). From 1996 to 2008, he also headed the cartographic department of GRID-Arendal (UNEP) in Norway. Since 2006, he has been dedicated to research projects exploring new forms of cartographic expression (radical, sensorial, and experimental mapping) participating in international carto-artistic events. He co-founded Visionscarto.net with Philippe Rivière. In 2017, he joined the Crosslocations program at the University of Helsinki and the Territories of Urban Extension program at ETH Zurich. Since 2020, he has been a researcher and associate lecturer at Wageningen University in the Netherlands. Author of thousands of cartographic documents, his style and distinctive touch reflect his research in graphic semiology and geometry.

Rueß, Simone is a visual artist who teaches art and design at various universities, e.g. at the Goethe University in Frankfurt a. M. and the University of Wuppertal, and is an associate member of the Collaborative Research Centre 'Re-Figuration of Spaces' (CRC 1265) at the Technische Universität Berlin. She studied at the Academy of Fine Arts in Stuttgart and was a DAAD postgraduate at the Academy of Fine Arts in Warsaw, followed by a three-year research stay in the Polish capital, funded by the graduate scholarship of the state of Baden-Württemberg and the Kunstfonds Bonn. Furthermore she was a fellow i.a. at BASE Milano, Akademie Schloss Solitude and Cité Internationale des Arts, Paris. Her artistic research on urban structures, home-making and biographical narratives has been presented in various group and solo exhibitions, including Trafostacja Szczecin, Le Guern Gallery in Warsaw, Galerie Interface in Dijon, Kunstmuseum Ravensburg and La Triennale di Milano in Italy.

Saxena, Alder Keleman is an assistant professor in the Department of Anthropology and Program in Sustainable Communities at Northern Arizona University in Flagstaff, Arizona. She holds a PhD in Environmental Anthropology from Yale University, where she also completed master's degrees in Environmental Science and International Relations; and a BA in Biology and Latin American Studies from Scripps College. Her research explores human-plant relationships, and particularly the interdependencies linking small-scale farmers and agrobiodiversity conservation in Latin America. Her current book project, *Eating Agrobiodiversity: Risk and Rewards of Indigenous Foods in Plurinational Bolivia*, brings the more-than-human turn into dialogue with ethnobotanical scholarship, critically evaluating the potential for food-based movements to drive agrobiodiversity conservation and social justice simultaneously.

Schamun, Katrin, Dr., is an urban and media philosophy researcher. She has held positions as a teacher and researcher at Harvard University, the Academy of Fine Art Hangzhou and the Universität der Künste Berlin. She was researcher within the project 'Well-being in Socio-Spatial Contexts: Intersectional Perspectives on Children's Experiences at Non-School Learning Sites (WIKK*I)'. Her further research explores the experience of environments under contemporary conditions mediated by technologies, and the intersection of audio-visual media and society.

Sommer, Vivien, Dr., is a sociologist who researches the complex network of knowledge, space, and memory. Since November 2023, she has been the junior research group leader of the Emmy Noether Group 'The Socio-Spatial Memory of European Borders: Dispositives of Remembering and Forgetting' at the Leibniz-Institute for Research on Society and Space. She is also an associated member of the Collaborative Research Center 1265 'Re-Figuration of Spaces' at Technische Universität Berlin. An important part of her research is the development and adaptation of qualitative approaches for research on relational space. She focuses on theory-generating methods, discourse-analytical approaches, visual and multimodal methods, and mapping approaches.

Sperling, David, Ph.D. in Architecture, is professor of spatial studies at the Institute of Architecture and Urbanism of the University of São Paulo, Brazil. He is researcher of the Brazilian National Council for Scientific and Technological Development, developing the project Cartographies: Technopolitics and Geopoetics. He is Co-Head of the Center for Studies on Contemporary Spatialities of IAU-USP, and of the Art Science Technology Group of the Institute of Advanced Studies of the University of São Paulo, and Co-Coordinator of the *Ground Atlas* project (groundatlas.org). He studied architecture at the IAU-USP and did his doctorate in architecture (design, space and culture) at the FAU-USP. His research focuses on mappings and cartographies in the intersection of architectural studies, arts, geography, and philosophy.

Tsing, Anna L. is distinguished professor of Anthropology at the University of California, Santa Cruz, and professor of Anthropology at Aarhus University, Denmark. She is co-director of UC Santa Cruz's Center for Southeast Asian Coastal Interactions and a member of a team co-writing a book, currently entitled *Governance is Hardening: Where Land Meets Water in Southeast Asia*. The book follows up on the insights of the co-authors in *Feral Atlas* and the *Field Guide to the Patchy Anthropocene*; it explores the effect of building coastal infrastructure, especially wetland draining and concrete urban and harbor construction, on more-than-human lives and livelihoods.

van Dooren, Noël is a landscape architect. He operates between practice and academia as a designer, advisor, writer, and teacher, both independently and affiliated with an institute. His latest affiliations were as an advisor of the government concerning spatial affairs for the province of South Holland (NL) (2020–2024) and as researcher at Aeres University of Applied Sciences Almere (2021–2020). From 2004 to 2009, he headed the landscape architecture programme at the Amsterdam Academy of Architecture, part of the Amsterdam School of the Arts. That led to his PhD research, *Drawing Time*, on the relation between representation, time, landscape, and design, which he defended in 2017. In lectures and workshops, he expands on that theme, for example at the technical universities of Munich, Cottbus, and Delft. He has published on the topic in both academic and professional journals.

Willis, Anne-Marie, PhD., is a writer, editor and educator on design and cultural theory who has taught at universities in Australia, Hong Kong and Egypt, including as professor of Design Theory at the German University in Cairo. She holds a visiting professorship in the School of Architecture at the University of Adelaide, and teaches design ethics and design history at the University of Tasmania. She has worked in design consultancy as Assistant Director of the EcoDesign Foundation, a non-profit organisation founded with Tony Fry, which pioneered sustainable design in Australia. Anne-Marie was co-founder and editor of Design Philosophy Papers, and has published five books, including *Illusions of Identity: The Art of Nation* and the *Design Philosophy Reader*. She is currently working on a book of essays on the relation between design and ontology, to be published in 2026.

Zhou, Feifei is the founder of multidisciplinary design studio terriStories, and adjunct assistant professor at Columbia University, Graduate School of Architecture, Planning and Preservation. Her work explores spatial, cultural, and ecological impacts of the industrialised built and natural environment. Using narrative-based spatial analysis, she collaborates extensively with social and natural scientists to translate empirical observations and scientific research into visual representations that aim to both clarify intricate more-than-human relations and open new questions.

Imprint

An imprint of Walter de Gruyter GmbH,
Berlin/Boston

Funding Statement
The research was funded by the German Research Foundation (Deutsche Forschungsgemeinschaft). Projektnummer 290045248 – SFB 1265.

Cover: Katja Gretzinger
Coordination: Mara Oppermann
Proofreading: Cecilia Tricker
Design: Katja Gretzinger
Lithography: Bild1Druck, Berlin

Printed in the European Union.

For questions about the General Product Safety Regulation please contact productsafety@degruyterbrill.com

Bibliographic information published by the Deutsche Nationalbibliothek
The Deutsche Nationalbibliothek lists this publication in the Deutsche Nationalbibliografie; detailed bibliographic data are available on the Internet at http://dnb.d-nb.de

jovis Verlag
Genthiner Straße 13
10785 Berlin
www.jovis.de

jovis books are available worldwide in select bookstores. Please contact your nearest bookseller or visit www.jovis.de for information concerning your local distribution.

ISBN 978-3-98612-250-8 (softcover)
ISBN 978-3-98612-252-2 (e-book)